MAKING THINGS WORK

SOLVING COMPLEX PROBLEMS
IN A COMPLEX WORLD

MAKING THINGS WORK

SOLVING COMPLEX PROBLEMS IN A COMPLEX WORLD

YANEER BAR-YAM

EDITING AND FORMATTING BY CHITRA RAMALINGAM,
LAURIE BURLINGAME, AND CHERRY OGATA

NECSI
◆
KNOWLEDGE PRESS

ISBN 0-9656328-2-2

123456789
First Printing,

NECSI ◆ Knowledge Press
http://www.necsi.org ◆ http://www.knowledgetoday.org

This book is dedicated to:

my family:
Zvi, Miriam, Naomi, Shlomiya, Yavni, Maayan, Taeer,
Sageet, Dani, Neoreet, Lehaveet,
Aureet's memory is with us;

Cherry, Debra and Luke;

and my colleagues, students and postdocs.

Table of Contents

PREFACE

In recent years the rapidly changing world around us has been raising concerns about the ability of people to cope with change. *Future Shock*, *The Ingenuity Gap*, and other books describe the difficulty of people living in our complex world. Complexity may seem overwhelming but it is not a bad thing. The complexity of the world is a mirror reflection of ourselves working together to make the world work. We, together, are becoming increasingly complex. The reason we can do this is that we work together in increasingly effective ways. We are connected to each other in ways that allow us to respond as teams and organizations. This enables us to do things we would not be able to do by ourselves, not just in terms of amount of effort but in terms of complexity. Complex tasks require complex organizations. When we are part of a complex team we find the world a remarkably comfortable place, because we can act effectively while being protected from the complexity of the world. This feeling is like the experience of a cell in a body, protected from the environment, and contributing to the organism function. Today civilization is the organism we are part of. We are in the midst of a remarkable transition from the individual to the group, organization, and even to global civilization as a functioning unit. While this is a mind bending transition, it is a transition of opportunity for creating a world that works for everybody, on the global level and on the level of each individual.

During this transition, we can be impressed by how the world works, or we can be frustrated by the parts of the world that do not work effectively. The comforts and conveniences of affluent living in a developed society are remarkable: the availability of food, clothing and shelter far beyond their necessity, the ease of communication and transportation, and the opportunities for professional fulfillment and entertainment. Even seemingly small matters, like making plans for a meeting tomorrow, next week, or next year, assumes that the world works—that we won't catch a disease and die suddenly, or be in a car accident, or be taken away by a gang, or killed by terrorists, or unjustly accused of a crime, or even that we will have enough food to survive until then. At the same time, this assumption is not true for everybody. There are many places in the world (and groups in every society) where the world doesn't work; where children starve, and violence and disorder reign. There are also surprising ways that even the best of places don't work. We learn from experience as well as the news that it is not uncommon for a person to go into a hospital and die (or be seriously hurt) needlessly from an error even though the people who work there are devoted to making us healthy. We care that our children will develop well and achieve much, but cannot assume that they are receiving the best education, or even a reasonable one, though teachers and schools are doing their best to provide it. Terrorists can strike, and people may die or be injured from gang violence and other crime, despite military and police protection. When we consider our own actions, we find that it is difficult to ensure that what we do will be constructive both for ourselves and for others. Too many of our efforts fail or are dissipated in the manifold events that surround us. Indeed, it often seems that we have no effect, or that the effect may be counter to our good intentions.

The study of complex systems enables us to frame, understand and respond to these feelings and observations. One of the most profound results of complex systems research is that when systems are highly complex, individuals matter. It also helps to reveal the actions that should be taken in our complex context to improve our world for ourselves and others.

What should we do when things don't work? This book will respond to this question using results from the scientific study of complex systems. Everything we do and everything that works and doesn't work around us is embedded in the complex system of our social context and our society. Having devoted my attention to the effort to understand what makes complex systems work and not work, it is a great pleasure for me to have the opportunity to share what I have learned.

The problems that we will discuss in this book are important, but since they arise in the context of a complex society, I have little doubt that we will overcome the problems and progress to the next level of individual and societal accomplishment and well being. Though in everyday life we are bogged down by details and difficulties, we shouldn't lose sight of the contribution of each of us to making things work.

Part of understanding the world around us as a complex system is recognizing the way our actions involve other people. For me, things work because of the remarkable abilities of the people around me and their devotion to this effort. I wish to acknowledge their contributions to the creation of this book and the work that led to it, including those who support and enable the work to be done. This is a collective effort.

I wish to mention those who have been particularly connected with this work and the activities that led to it in recent years.

First among these is my father, Zvi Bar-Yam, who has enabled all of this to take place in many ways, he has been a partner in all that I have done. My mother, Miriam Bar-Yam, has enabled him to act and has helped directly in many ways. I have been taught by my parents a devotion to contribution to society. I learned from them (before I learned from the study of complex systems) that each person can contribute to improving the world and to helping others.

There are many who contributed to the writing and editing of this book, these include members of my family, particularly Zvi, Miriam, Naomi, and Shlomiya. Chitra Ramalingam worked to compile diverse materials into book form and did the first editing, Laurie Burlingame did much of a second round of editing, Cherry Ogata formatted the manuscript and finalized the figures, and Gregory Wolfe worked on the index. David Roberson suggested the primary title.

Helpful comments were received from draft readers: Bob Arnold, Phyllis Bromberg, Hans-Peter Brunner, Robert Davenport, Osvaldo and Sylvie Golijov, Richard Grossman, Diego and Rosie Jaramillo, Tom Siegfried, Peter Tallack, Joan Wells, and Norman Zarsky.

My sister Sageet has been working to bring complex systems to science and society, and others have come to share the effort. Among the people who have become members of the team are Cherry Ogata, Debra Gorfine, Luke Evans, and Isabel Cunha-Vasconcelos. Key additional help has been provided by Don Glazer, Larry Kletter, John Imbergamo, and Peter DeIeso.

I would like to thank my colleagues and collaborators, especially: Michel Baranger, Michael Benari, Bruce Boghosian, Dan Braha, Hans-Peter Brunner, Charles Cantor, Jeff Cares, Teo Dagi, Alice Davidson, Terrence Deacon, John Dickman, Meghan Dierks, Kerry Emanuel, Irving Epstein, Roger Frye, Frank Funderburk, Marcus de Aguiar, William Gelbart, Edward Goldberg, Ary Goldberger, Charles Goodnight, Helen Harte, Ernest

Hartmann, Sui Huang, Don Ingber, Jim Hansen, Michael Jacobson, Jerome Kagan, James Kaput, Mehran Kardar, Stuart Kauffman, Les Kaufman, Mark Klein, Mark Kon, Mike Kuras, Blake LeBaron, Jerome Lettvin, Seth Lloyd, Norman Margolus, David Meyer, Bill Mills, Ali Minai, Mai Nguyen, Doug Norman, Alan Perelson, Tom Petzinger, Yitzhak Rabin, David Roberson, Daniel Rothman, Larry Rudolph, Peter Senge, Mark S. Smith, Temple Smith, John Sterman, James Stock, Jerry Sussman, Stuart Pimm, Hiroshi Tasaka, Gunter Wagner, Michael Werman, David S. Wilson, Tom Toffoli, Norm Margolus, Kosta Tsipis, John Wakeley, Michael Werman, Sheldon White, Bob Wiebe, and Uri Wilensky.

The contributions of students and postdocs who have worked with me are greatly appreciated: Ed Addison, Gavin Crooks, Benjamin de Bivort, Raissa D'Souza, Peyman Faratin, Speranta Gheorghiu-Svirschevski, Daniel Goldman, May Lim, Richard Metzler, Boris Ostrovsky, Darren Pierre, Erik Rauch, Jason Redi, Kathleen Rhoades, Daniel Rosenbloom, Reza Sadr-Lahijany, Hiroki Sayama, Finley Shapiro, Ben Shargel, Muneichi Shibata, Mark A. Smith, Harsh Vardhan, Justin Werfel, and Sanith Wijesinghe.

I have also benefited from the many outstanding discussions, projects and presentations of students in my courses at BU, MIT and NECSI, and from the many participants in NECSI programs, discussion groups, workshops, lectures and seminars, and speakers and participants in the International Conference on Complex Systems.

I am indebted to my teachers, from elementary school through college and especially my PhD advisor John Joannopoulos.

I would like to recognize the importance to our work of support from grants from government agencies, corporate and institutional donors, and particularly from NSF, NIH, MITRE, AFRL, SSG, CMS, CDC, Sophia-Bank, Pan Agora Asset Management, Dean LeBaron, Edgar Peters, and Hiroshi Tasaka.

The many collaborators and supporters have not reviewed or endorsed the content of this book. If there are errors, they reflect my limitations, as well as the ongoing opportunity to learn.

Yaneer Bar-Yam
Newton, 10/28/04

תושלב"ע

Overview:

Making Things Work

Today we often describe the world around us as highly complex. Complexity manifests in everything from individual relationships to corporate challenges to concerns about the human condition and global welfare. As a global community, we are in the middle of a transition from the industrial to the information age, and this transformation is reflected and rereflected in everything around us. The amount of information that is flowing and the rate of change of society are both aspects of the growing complexity of our existence. As individuals, we have a hard time coping with all the information and change. In some sense more importantly, our society is also having difficulty coping with its own changes.

Our economic and social institutions, that we rely upon at critical times of our lives, including the health and education systems, are changing, not always gracefully, to meet the new challenges. Professional activities, from corporate management to systems engineering, require new approaches, insights and skills. Global concerns, such as environmental destruction and poverty—in developed and undeveloped nations—are becoming more pressing as these changes take place.

Despite major efforts to identify the solutions to these problems, they are often obscure and hidden from us. Even when we think we are making progress, the solutions we think of today may cause us more problems tomorrow. This is because complex problems do not lend themselves to easy solutions. Any action may have hidden effects that cause matters to become worse and the whole strategy we are using may be moving things in the wrong direction. Complex problems are the problems that persist—the problems that bounce back and continue to haunt us. People often go through a series of stages in dealing with such problems—from believing they are beyond hope, to galvanizing collective efforts of many people and dollars to address the problem, to despair, retreat, and rationalization. The progress made seems miniscule compared to the effort and resources expended. Even with all of the modern technological advances, it is easy to become pessimistic about the world today. There is hope, however, in the recognition that people can solve very complex problems when they work together effectively. Unfortunately, this is generally not how we respond when there are problems. We don't always realize the ability that we have when we work together. We tend to assign blame or responsibility to one individual.

When a problem arises, there is a strong tendency to try to figure out who is responsible for it. Someone should be fired; someone should pay; someone should be punished. Today, in an important step forward, there is an increasing tendency to use a "systems perspective," to recognize that many factors may be responsible—too many to be identified individually. This is the notion that the "system" is at fault. Unfortunately, this step forward is not sufficient for knowing how to address the problem. Most of the time, when we hear someone say "it's the system," it means that the speaker is abandoning all efforts to fix the problem. The complexities of the system are daunting and a clear fix is not apparent. These feelings lead to total paralysis of action! The purpose of this book is to provide tools for thinking about how to fix problems that have to do with the system.

What do people do today when they don't understand "the system?" They try to assign responsibility to someone to fix the problem, to oversee "the system," to coordinate and control what is happening. It is time we recognized that "the system" is how we work together. When we don't work together effectively putting someone in charge by its very nature often makes things worse, rather than better, because no one person can understand "the system" well enough to be responsible. We need to learn how to improve the way we work together, to improve "the system" without putting someone in charge, in order to make things work.

Before we can explain how system problems arise and can be fixed, we have to understand something about how systems work. This is where science can help. For many years there has been a sense that chaos and complexity, promising new areas of scientific inquiry, have something fundamental to tell us about the world in which we live. James Gleick's classic book *Chaos: Making a New Science* (1987) and many other books in later years have raised popular awareness of these directions of research. Much of the focus has been on recognizing the intrinsic unpredictability of nature, and—by extension—of society. However, beyond the fascinating applications to turbulence, meteorology, and other complex problems in the natural world, complex systems science has more to tell us about the world—including human beings and their interactions—than just that it is unpredictable. The concepts that help us in general to analyze complex systems can guide us in confronting our pressing problems.

This book provides a description of concepts as they have been developed in the scientific study of complex systems, but here they're directed at solving the complex problems of our world. We'll discuss the U.S. military's successes with integrating ideas about complexity into their practices (and failures when it doesn't use them), the difficulties that have plagued the effectiveness of the health care and education systems, and new ideas regarding socioeconomic development in the third world. We'll also discuss large engineering projects like the efforts to modernize the aging air traffic control system. In all of these areas, complex systems concepts can provide new insights about how to approach solving these difficult and deeply rooted problems.

There have been many books on management that evoke useful complex systems concepts like self-organization and networks. However, many of these treatments do not consider important trade-offs and paradoxes. The general sense these days is that highly connected networks are always a good idea and that self-organization can be trusted to achieve any objective. Words like "integrated" and "interconnected" are used as if they are synonymous with "good." These assertions are too strong. They misrepresent the most important insights that complex systems science can bring to solving real world problems. In general, there is no one-way to solve all complex problems. In management terminology, there are no "best practices" that apply to all cases. The solution to a problem has to be related to the type or structure of the particular problem.

Still, there are methods of thinking about complex problems that are often, if not always, useful. These have to do with relating the nature of the problem to the nature of the solution, a kind of yin-yang complementarity. Through many years in the field of complex systems, my colleagues and I have developed an approach to understanding complex systems based on a few fundamental ideas:

- the mechanisms of collective behavior (patterns),
- a multiscale perspective (the way different observers describe a system),
- the evolutionary process that creates complex systems, and
- the nature of purposive or goal-directed behavior.

Whether we use them to talk about biological molecules or corporations, these interwoven approaches help us classify complex systems, recognize their functional capabilities and develop a context in which their strengths and weaknesses can be evaluated.

The major topics covered in this book include:

The Health Care/Medical System. The health care system is struggling with the dichotomy of large-scale undifferentiated financial flows and highly complex medical treatments that require individual doctors to make careful decisions about the treatment of individual patients. Modern efforts to lower costs, by creating efficiency through industrial era methods, are incompatible with complex specialized treatment and are leading to increased medical errors and decreased quality of care. More specifically, the problem of medical errors (as with errors in other systems) arises because the system is not well designed for the high complexity tasks it performs.

Education System. The traditional approach to solving social problems with large scale forces is now being used to tackle the complex problems of the education system. While the failure of many schools to provide quality education is real, the current approach to solving the problem, through standardized testing, is anachronistic. Standardized testing for student, school, school system, and curriculum evaluation is an industrial era approach of mass production for uniform products. The complex society of the information age needs diverse skills and, needless to say, individual desires and talents will not be fulfilled through mass production of students with standardized, and ultimately lowest common capabilities.

Corporate management. The transformation of management to address highly complex challenges in the environment began in earnest in the 1980s. Today the inability of conventional hierarchical control and the need to understand distributed control, self-organization and networks is increasingly apparent. The recent financial fiascoes at high profile companies are the latest evidence of the disassociation of management from corporate function. In discussing information age companies, it is important to recognize that there is no one set of "best practices," the primary organizational principle is a matching of the system structure to the environment and function it performs. Recognizing key characteristics of the functional demands on the system can guide choices that are made about organizational structure and information flows. Moreover, the primary mechanism for organizational learning and change is through evolutionary processes. Finally, the attention to personal and corporation-wide networks is an abstraction of the centrality of relationships (relatedness).

International Development. A functioning economy is a highly complex organization. Anticipating, designing, or planning the behavior of such a system does not work. Detailed planning of interventions for desired outcomes is still the main approach of development agencies like the World Bank. Almost any large scale intervention is likely to be destabilizing because such interventions are fundamentally incompatible with existing, as well as desired, fine scale socio-economic interdependencies. Since complex systems exist in relation to their environment, there is a self-consistent relationship between each country's economy and its environment (natural environment within its borders as well as natural and human environment outside them) which must be recognized in development efforts. Moreover, any intervention becomes entangled with system functioning, so that the goal of promoting effective and independent functioning of a country by direct intervention is paradoxical. Insights into pattern formation and evolutionary dynamics are needed to overcome these obstacles to social improvement efforts.

Military. The military has learned important lessons about complex warfare from Vietnam and other recent military conflicts. The use of a radically different military strategy in Afghanistan from that in the Gulf War in 1991 manifests this understanding. Recognizing the complexity of terrain, enemy forces, political contexts and its impact on goal setting, strategy, operations and tactics have been key to effective action. These lessons are being incorporated into military doctrine, technological innovation and modernization programs. In many contexts, the military has explicitly stated its recognition of the role of insights from the general study of complex systems. Unfortunately the relevance of many of these lessons was either not recognized or ignored in planning the current war in Iraq.

Engineering. In the mid 1990s, after 12 years of effort and a cost of \$3–6 billion, a project to redesign the U.S. air traffic control system was abandoned without replacing a single part. The existing system, developed forty years earlier in the 1950s, was still using vacuum tubes. This is but one example of many failed systems engineering projects. The difficulties arise even when outstanding engineers use well established state-of-the-art methods. The problem of design and construction of new systems, or replacements for highly complex systems, that are critical for the government (military or civilian) as well as for major corporations, has repeatedly been found and can be formally proven to be beyond the capability of traditional systems engineering practice. Today, we are developing a new strategy for complex systems engineering based upon an understanding of how complex systems arise in nature.

International terrorism. The challenge of modern terrorism and asymmetric warfare should be understood from the perspective of global change. The underlying conflict arises primarily from a global differentiation between cultures. The "clash of civilizations" is not a desire for conquest, but rather a process of clarifying the boundaries between distinct locally incompatible cultural systems. Accelerating the establishment of well-defined boundaries seems to be the best strategy to global peace. It is also a key to reducing the ability of extremist groups to conduct terror operations. The elimination of specific individuals is less likely to be effective in reducing the overall threat of terrorism.

Some of the fundamental ideas in this book have been discussed by others, but many have not. They are based on my own research in complex systems. The applications to real world problems are from my experience as president of the New England Complex Systems Institute. In this capacity, I have been asked to develop educational programs and teach complex systems ideas to military planners and the World Bank. I have also developed an executive education program attended by executives from diverse industries, but particularly from health care. I have been asked to consider problems in the education system and in systems engineering.

The exploration of these problems has reinforced my perspective that research in complex systems is remarkable in being both highly applied and practical even as it is fundamental. The opportunities for developing an understanding of complex systems through scientific research are only beginning to be explored. It is an exciting time as we step out of the traditional areas of research to discover that we now have new approaches and insights that can help us answer some of the key problems we face in the world around us.

Some of the challenges discussed in this book, such as health care and education reform, have been described as "crises" for years now. Despite vast investments of money and attention most major efforts are ineffective and stalling. In some of the other areas, like engineering and international development, there's a general sense in the air that methods of approach are changing, but a comprehensive set of tools to analyze and constructively engage with these problems is still lacking. In the military, they've been studying complexity for years and when the lessons are used you can really see the difference.

The discussion of each of these problems in this book serves as a case study of the application of complex systems concepts to understanding systems and solving problems. These examples also illustrate how we can use complex systems ideas to understand the world around us, including our own day-to-day experiences.

Developing the ability to use a complex systems perspective requires new patterns of thinking. In the first section of this book some of the key complex systems ideas are described. These ideas—like emergence and interdependence—have to do with relationships between parts of a system and how these relationships lead to the behavior of the system. After all, society works because of how people interact with and relate to each other, not how each person acts separately. The results of the interactions between people are patterns of behavior. We will look at how patterns can arise from interactions without someone putting the parts of the pattern in place by telling each person what to do. Using our understanding of how neurons interact in the brain, we will show how the pattern of behavior can be made to serve a purpose. We will find that the type of pattern that arises can be related to how the system is organized—who can interact with whom. We will look more generally at the set of things a system can do, and how this set of actions is related to how it is organized. Some organizations are good at doing complex tasks, and some are not. Perhaps not surprisingly, centrally controlled or hierarchical organizations are not capable of highly complex tasks. This means that we have to figure out how to make distributed/networked organizations if we want to solve complex problems. Finally, we learn about evolution, how really complex systems (including distributed/networked organizations) can form and be effective without being planned (which is crucial because planning them doesn't work!). Counter to how evolution is usually discussed, it is not just about competition, it is always about both competition and cooperation. Competition and cooperation work together at different levels of organization, just as in team sports where players learn to cooperate because of team competition. Making an effective organization is making a successful team.

Through the discussion of these ideas we can explain how individuals can work together and solve complex problems: first how individual acts combine together, second what makes them effective, and third how this effectiveness can arise and improve over time.

The second part of the book applies these ideas to the real-world problems mentioned above showing how we can organize ourselves to solve complex problems in a complex world.

PART I:

CONCEPTS

Chapter 1

Parts, Wholes, and Relationships

Parts, wholes, and relationships

Scientists look at something and want to understand what it does, and how it does it. One of the key observations about the world is that everything is made up of parts. So reasonably enough, scientists, trying to figure out what the object does, work to figure out what the parts do. When we look at one of the parts, we realize that it too is made up of parts. The next step, therefore, is to look at the parts that make up the part to figure out what they do. This continual breaking down of parts into their component parts progresses until we forget what it was that we were trying to do in the first place!

Consider for a moment all of the levels of parts that make up the human body: The body is formed out of nine organ systems, which are formed from a variety of organs, which in turn are formed from tissues, which are formed of cells, which are formed of organelles, which are formed of molecules, which are formed of atoms, which are ultimately formed of elementary particles! The same types of molecules form all biological systems, including the human body, and the same types of particles form all matter, both living and nonliving. These are powerful and surprising insights that today scientists take for granted. Trees and rocks are made of the same building blocks. Physicists take this for granted, and consider the study of elementary particles to be the study of all of nature. People and trees are made of the same building blocks. Biologists take this for granted, and consider the study of biological molecules to be the study of all life. What is left out of this approach is the relationships that exist between the parts of an object.[1] There is no doubt that science has made great progress by taking things apart, but it has become increasingly clear that many important questions can only be addressed by thinking more carefully about relationships between and amongst the parts. Indeed, one of the main difficulties in answering questions or solving problems—any kind of problem—is that we think the problem is in the parts, when it is really in the relationships between them. This book will explain why an understanding of the underlying relationships between the parts is so important.

Scientists generally think that the parts are universal, but that the way the parts work together is specific to each system. Those of us who have been exploring complex systems, though, are beginning to realize that how parts work together can also be studied in general. By doing so we gain insights into every kind of system that exists, from physical systems like the weather, to biological systems like the human brain, to social systems like the American economy.

"Complex Systems" is a new approach to science, which studies how relationships between parts give rise to the collective behaviors of a system and how the system interacts and forms relationships with its environment.[2] Social systems arise (in part) out of relationships between people, the brain's behaviors result from relationships between neurons, molecules are formed out of relationships between atoms, and weather patterns are formed because of relationships between air flows. Social systems, the human brain, molecules and weather patterns are examples of complex systems. Studying complex systems cuts across all disciplines of science, as well as engineering, management, and medicine. It is also relevant to the humanities: art, history, and literature. It focuses on certain questions about relationships and how they make collections of parts into wholes. These questions are relevant to all systems that we care about.

There are many advances that have made complex systems an exciting area of research today. It is impossible for me to discuss all of them in this book, but I can give you a taste that will, I hope, invite further inquiry. In this overview I will introduce the concepts of emergence and interdependence. Then, in the next six chapters, I will describe three interrelated approaches to the modern study of complex systems (in the second, fourth and sixth chapters), and pair each of them with an interesting extension and application (in the third, fifth and seventh chapters).

The second chapter discusses how self-organized patterns of behavior arise from interactions between parts. For example, simple models of influences between people can be used to understand seemingly mysterious phenomena like fads and panics. In the third chapter we will look at how models of influences in networks can be used to study more complex patterns of social behavior, or the patterns of behavior of neurons in the brain. Using these patterns, the network structure of the brain can be related to abstract properties of the mind. To illustrate how this is possible, we will discuss how the structure of the brain is related to human creativity. Similar ideas apply to social networks, revealing how we can think about people working together.

The fourth chapter is about how we can think about describing complex systems and how our intuitive ideas about complexity can be made more precise. We will find that complexity and scale are balanced against each other. Here, the word scale is used just as it is used in phrases like "economies of scale" or "scale of operation," referring to the scale of the activity that is taking place. These ideas are also related to exploring the space of possibilities—the possible patterns that can happen, not just the one that is happening. In the fifth chapter we will show that the balance of scale and complexity helps us understand how social systems are organized and how historical changes in society are leading to a networked global civilization.

The sixth chapter discusses evolution and how making many small incremental changes can be an effective way to make complex systems. The classic way to think about evolution is to consider the competition for survival as giving rise to these changes. It is important to realize, however, that this picture is not really complete. Cooperation and competition always work together and an essential part of evolution is the formation of groups and collective, cooperative behaviors. In the seventh chapter we will illustrate how cooperative behaviors are what enable competition and competition is what enables cooperative behaviors. These are natural concepts in the context of a discussion of team sports.

As a first step into the study of complex systems, we will begin by describing two concepts: emergence and interdependence.

Figure 1.1: Forest and hills: the large-scale view.

Figure 1.2: Trees, animals, plants: the smaller-scale view.

Emergence[3]

In Figure 1.1 we see a forest on hills. In Figure 1.2 we see trees, plants and animals. The forest on the hills is made up of many trees, animals and other plants. What we see in Figure 1.2 could be a close-up view of one part of the forest in Figure 1.1. You've probably heard the old saying, "You can't see the forest for the trees." This statement captures a fundamental insight about complex systems: when you focus on the small-scale details of a system or a situation—the growth and development of individual trees, what plants certain animals consume—you run the risk of missing the larger picture. If your camera zooms out from the small scale to the large scale, you suddenly recognize that the forest has its own higher-level behaviors. Fires and regrowth, for example, are part of the natural behavior of a forest.

Most of science today has focused on the trees, studying the parts of a system, usually in isolation and ignoring higher-level phenomena. This approach creates barriers to the effective understanding of complex systems. However, focusing solely on the large scale view is also not adequate. Anything the forest does as a whole is made up of the behaviors of the trees and other plants and animals. Forest behaviors are *collective*: they are what parts of the system do together. Indeed, in many cases the behaviors are such that trees would not (or even could not) do them by themselves.

In conventional views the observer considers either the trees or the forest. Those whose zoom lens is focused on the trees consider the details to be essential and do not see the patterns that arise when considering trees in the context of the forest. Those who view the system from farther away, to observe the forest, do not see the details. When one can shift back and forth between seeing the trees and the forest one also sees which aspects of the trees are relevant to the description of the forest. Understanding this relationship in general is the study of emergence. *Emergence* refers to the relationship between the details of a system and the larger view. Emergence does not emphasize the primary importance of the details or of the larger view; it is concerned with the relationship between the two. Specifically, emergence seeks to discover: Which details are important for the larger view, and which are not? How do collective properties arise from the properties of parts? How does behavior at a larger scale of the system arise from the detailed structure, behavior and relationships on a finer scale?

When we think about emergence, we should, in our mind's eye, be moving between different perspectives. We see the trees and the forest at the same time. We see the way the trees and the forest are related to each other. To see in both these views we have to be able to see details, but also ignore details. The trick is to know which of the many details we observe in the trees are important to the overall behavior of the forest.

Interdependence[4]

The study of complex systems also helps us recognize and understand indirect effects. Problems that are difficult to solve using traditional approaches are often hard because the causes and effects are not obviously related. Pushing on a complex system "here" often has effects "over there" because the parts are *interdependent*. This has become more and more apparent in our efforts to solve societal problems or avoid ecological disasters that were caused by our own actions. The field of complex systems provides a number of sophisticated tools to address these difficulties, from concepts that help us think about these systems, analytical methods for studying these systems in greater depth, and computer-based techniques for describing, modeling or simulating these systems.

The first step, however, is just to begin thinking about how parts of a system affect each other. If we take one part of the system away, how will this part be affected and how will the rest of the system be affected? Sometimes the effect is small and sometimes the effect is large; sometimes there are many effects and sometimes only a few. Consider, for example, the effect of taking away a part from each of the following three items, a material (i.e., a piece of metal or a liquid), a plant, and an animal.

When you remove a small piece of a material (Figure 1.3), the internal properties of both the part and the whole are basically unchanged. The piece doesn't notice that it's been removed and neither does the rest of the material.

If you take a small part of a plant away (Figure 1.4), like a branch or some roots, the plant will typically continue to grow more or less the way it would otherwise. There are exceptions for certain crucial parts, like cutting a lateral part of the trunk, but generally the plant will not be strongly affected. On the other hand, the part of the plant that is cut away—the leaf or branch or piece of root—is very strongly affected. It will generally die unless it is placed in very special conditions.

Now imagine taking a small piece out of an animal (Figure 1.5). Ouch! We are not just talking about removing part of the wool of the sheep in the figure. Taking a part of the animal away will have devastating effects on both the part and the rest of the animal.

These three examples (material, plant and animal) exhibit very different kinds of interdependence. Recognizing that these different behaviors exist is an important part of characterizing all of the systems that interest us. Consider the family or organization you are part of and try to answer the following questions: How strong are the dependencies between the parts? What would happen if a part were taken away? Does it matter which part? These questions are key for understanding the system and how actions we could take might affect the functioning of the system. Just by asking these questions when we think about our world, we are taking an important first step towards understanding relationships and relatedness.

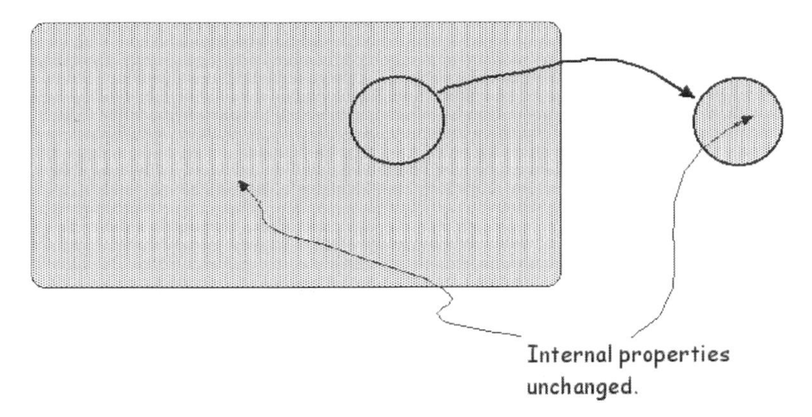

Figure 1.3: Removing a part of an inert material (liquid, solid, or gas).

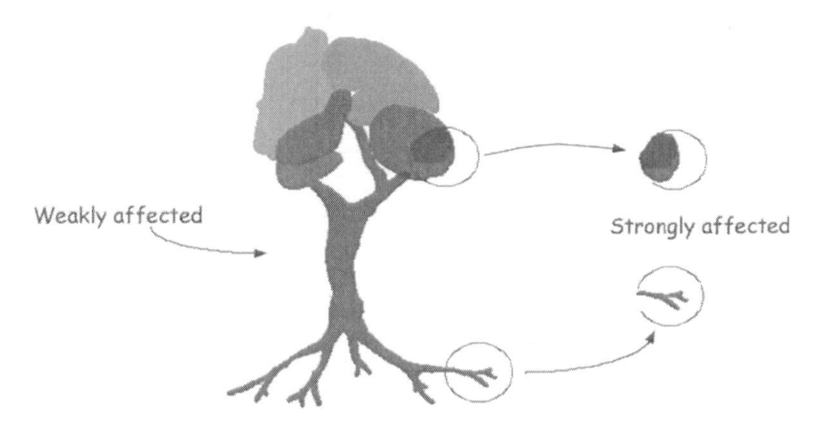

Figure 1.4: Removing a part of a plant.

Figure 1.5: Removing a part of an animal.

Chapter 2

Patterns

What is pattern formation?

When a car is built on the assembly line, each part is carefully placed in a particular location in order to create a structure that can accomplish a specific task. When an artist paints, he places each patch of paint in a particular spot to make the design he desires. Sometimes in nature, patterns form without anyone putting each part in a particular place. The pattern seems to develop all by itself: it *self-organizes*. Sometimes these patterns are regular, like ripples of sand on a beach or in the desert (Figure 2.1).

One of the most remarkable of all patterns is the human body itself. As with all animals, the human body grows from a single cell by a process

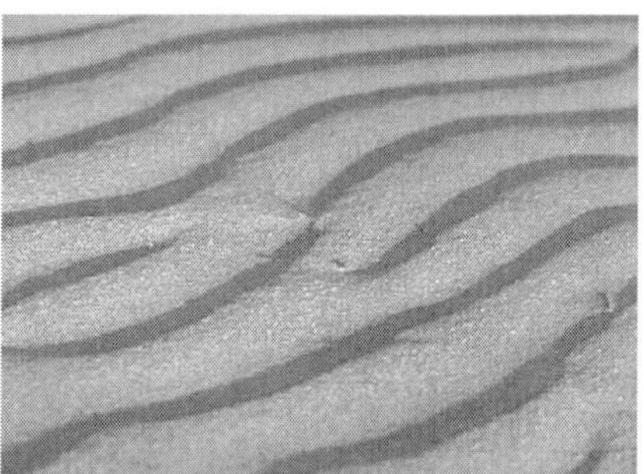

Figure 2.1: Examples of regular patterns that occur in nature

of *development* (Figure 2.2). During development, some of the cells form the heart, while others form the liver or the bones. There is no agent that actually puts each part in place, and yet when the process is complete, the parts work together in an intricate fashion. How do the cells know where to go, or what form and function to take in each part of the body?

At one time it was thought that the first cell contained a miniature human being within it, a "homunculus," that simply grew in size to produce a

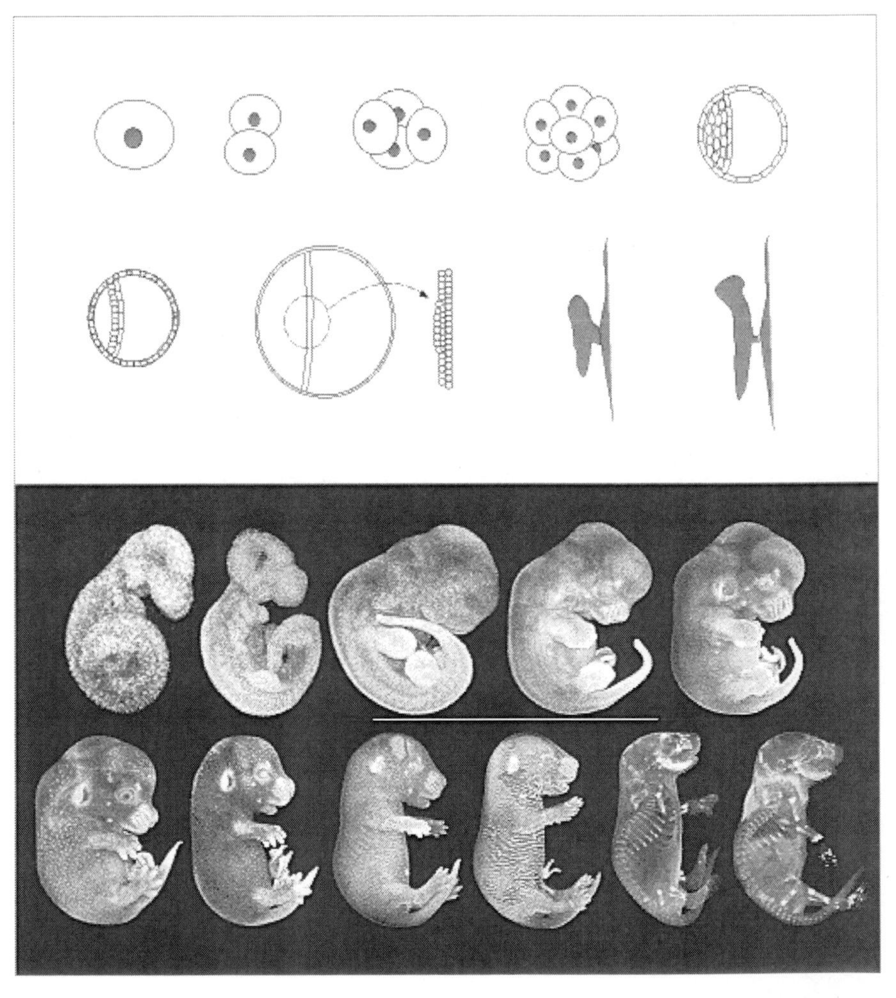

Figure 2.2: Mouse development. The first two rows are schematic drawings; the bottom two are images of developing mice. (Images courtesy of Brad Smith, Elwood Linney and the Center for In Vivo Microscopy at Duke University – A National Center for Research Resources, NIH])

mature human being. We now know that this is not correct. Still, there *is* something in that initial cell that in some sense "contains" the future human being: information, stored in DNA found in the nucleus of the cell. People often call DNA a "blueprint" but this is misleading—as much a mistake as the idea of a homunculus. A blueprint is a picture of the intended structure with each part shown in its place. The information in DNA isn't a picture of a human being. In some subtle way, a way we do not fully understand, the DNA tells the cell how it should "talk" to other cells. As they talk to each other, they form the structures of the body. Imagine giving instructions to a brick about how to talk to other bricks, then walking away and returning to find a house—with all of the windows, plumbing and electrical systems in place. It's not easy to imagine how this could be done, even if we did have bricks that could move around and morph into pipes, electrical wires, and insulation.

As scientists, we would like to understand how this self-organizing process takes place in nature. We would like to understand the mechanism by which patterns form and how the pattern that arises is determined. There are even wider implications than understanding how an embryo develops. Entire industries are based on the problem of how to bring intricate structures into being in a reliable and flexible way. If we could harness the natural process of pattern formation, it could revolutionize engineering and management. Imagine how different it would be if instead of specifying each of the parts of a system we want to build, we could specify a process that will create the system. This process would use the natural dynamics of the world to help us create what we would like to have.

One of the central lessons of complex systems is that external forces alone cannot explain how complex patterns form—including patterns of human behavior in economic and social systems. The interactions between people are crucial to understanding how fads and panics arise, or how stock market prices fluctuate from day to day. If you open the Wall Street Journal on any given day, you'll find statements explaining exactly why the market rose or fell on the previous day. However, these explanations are inadequate and often conflicting. Markets are self-organizing patterns and their behavior can only be described adequately if the interactions within them, among the buyers and sellers, are understood.

Examples of simple patterns[5]

In order to start thinking about the patterns we find in nature and in society we can begin by considering quite simple patterns. Consider a class of

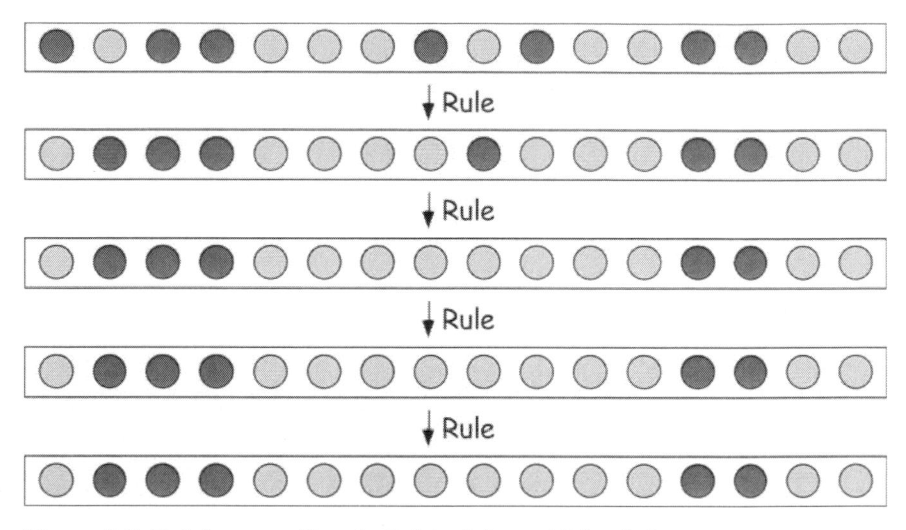

Figure 2.3: Pokémon or Beanie Babies? A model for fads.

kindergarten children at circle time. As children often do, these kids are talking with their neighbors and discussing which toy they want to buy: Pokémon® cards or Beanie Babies®. In Figure 2.3, the top row shows what each child wants to buy when they first come to class, indicated as black (Pokémon cards) or gray (Beanie Babies). (They are shown in a row for simplicity, but assume that this row is wrapped into a circle, so that the children on the ends of the row are sitting next to each other.) To see what happens as they talk to each other, focus on one of the children at a time to see what he or she is doing. For example, the first gray dot on the left is a boy who wants to buy Beanie Babies at first. However, he discovers to his dismay that his friends on either side both want to buy Pokémon cards. In order to have someone to play with, he changes his mind and switches to Pokémon cards.

Put yourself in the place of one of these children. As long as one or both of your neighbors is going to buy the same thing you are, you'll be satisfied with your choice. On the other hand, if you discover that nobody next to you wants the same thing, you'd probably have doubts. Is it really a good idea to get a toy that you'll have to play with all by yourself? You'd better switch to the other. If all of the children make their decisions according to this rule, we'll end up with a new configuration, shown in the second row of Figure 2.3.

Notice the gray dot near the middle. This is a girl who started out wanting to buy a Beanie Baby and changed her mind because of her neighbors.

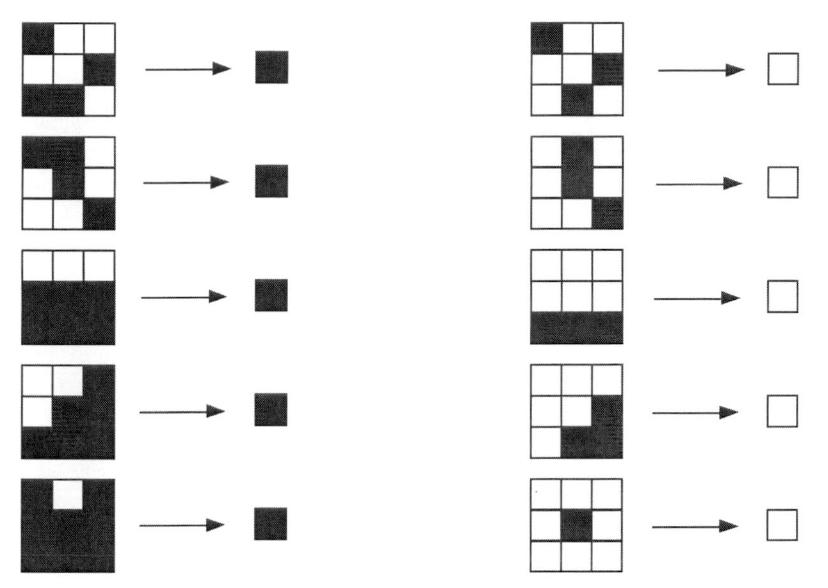

Figure 2.4: A model for panic, based on local interactions. Black squares represent panicky people; white squares represent calm ones. In the panels on the left, the person in the center will panic; on the right, the person in the center will remain or become calm.

Oops! A few minutes later, her neighbors have changed their own minds too. Hopefully she will talk with them again and switch back before going to the store. This is shown in the next row, which shows what the children want to buy a few minutes later, when they've checked with their neighbors again.

Notice what happens to the pattern of decisions in the circle after many repetitions; the children form patches of Pokémon or Beanie Baby buyers. Once they form, these patches are stable over time. The dynamics of this model could be used to represent other situations, like voting in a two-party presidential election, or buying and selling in a stock market. As people talk to each other, their preferences change.

We could also use a model like this to think about the spread of panic throughout a room. Consider people sitting in a crowded auditorium. They are aware of the people immediately surrounding them, in the same row and in the rows in front of and behind them. If there are enough people panicking around an individual, he will tend to panic too, even if he was calm. Figure 2.4 shows several cases of what a particular individual might do as a result of the influence of his neighbors. Each 3x3 panel shows the panicky people as black squares and the non-panicky people as white

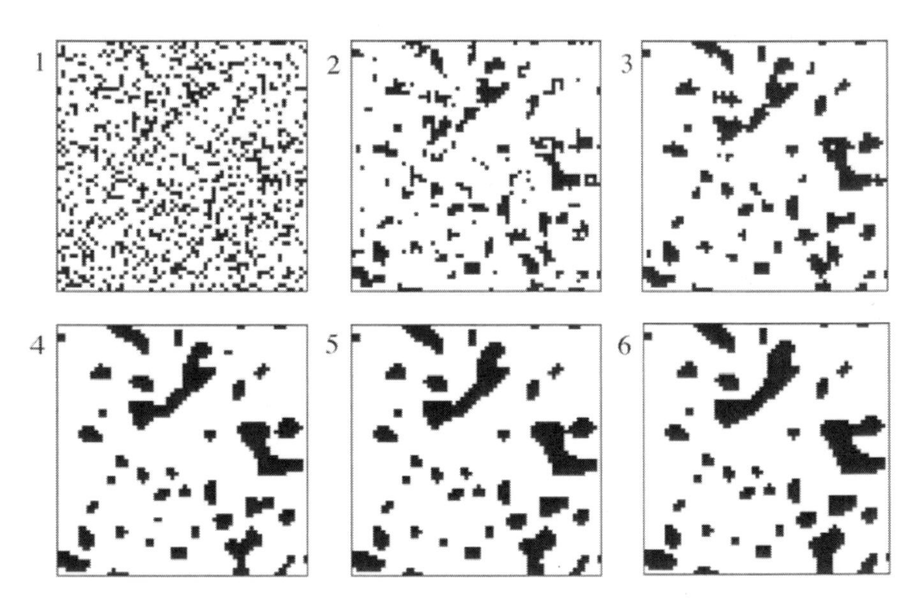

Figure 2.5: Model of panic in a crowded auditorium: Six repetitions of the panic rule.

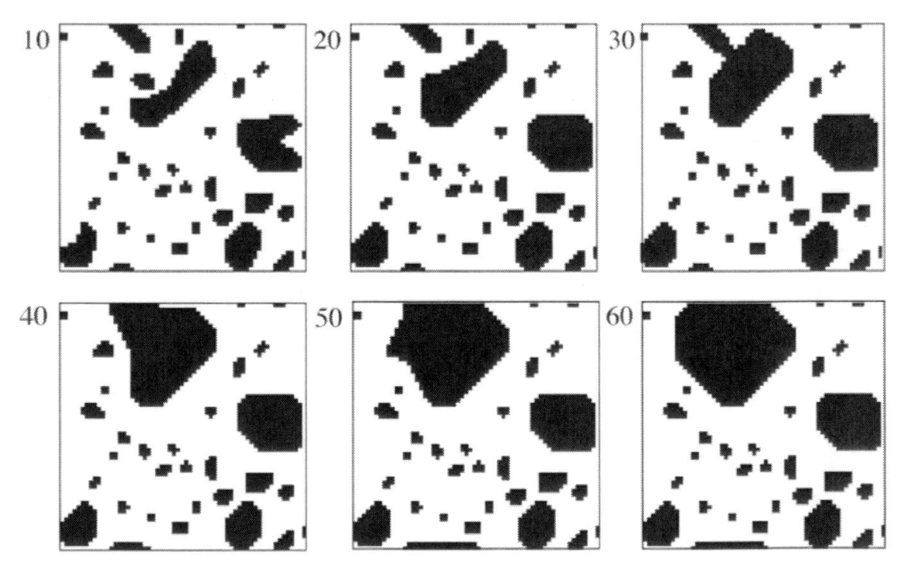

Figure 2.6: Continuation of the panic in an auditorium. There are ten intervals of time between each frame.

squares. If four of the people (count also the middle one) are panicky, then a few moments later the middle one will become panicky. If there are three or fewer, then he doesn't panic. Even if he himself was initially panicky, he calms down.

Now imagine someone enters a crowded space and yells "Fire!!" Depending on how he yells, some fraction of the people in the room will tend to panic. In regions of the auditorium where there are more of these people, the panic will spread. In regions where there are fewer it will tend to disappear. Will the panic spread throughout the room or not? Figure 2.5 shows a simulation of this rule in an auditorium. Each small cell represents a person and white/black corresponds to non-panic/panic. The first six frames are the first six intervals of time.

The next six frames in Figure 2.6 are separated by ten intervals of time. The last one shown is a stable pattern that would not change in subsequent frames.

Over the first few updates, the random arrangement of dots resolves into areas of panic. Isolated panickers calm down and regions of higher density become the areas of panic. Then, over a longer period of time, the panicking areas grow and reach a stable configuration. We can try this from a different initial arrangement of panickers. In some cases the panicking areas grow until they combine and fill the entire space.

It turns out that for this panic rule, in a space this size, the panic will grow to cover the space only if the room starts with more than a quarter of the people panicked (black). If it begins with less than this, the panic will stay in isolated patches, as it does in the figure. We can generalize this into a model of fads, mobs and hysteria, especially because it illustrates an important point—the existence of *transitions* in collective patterns of behavior. Sometimes behaviors feed on each other and grow to involve many people, and sometimes they don't. Understanding exactly what leads to the difference can be quite hard, but it definitely has to do with the interactions between people, the conditions under which the people are interacting with each other, and the triggering influence (if any).

This model is too simple to really explain how fads and trends work because people are more complicated than this model would suggest. For example, let's follow the kindergarten children as they grow up and go to high school. At this age, teenagers want to do the same thing as their friends, but they don't want to do what many other teenagers are doing. If anything, they will tend to do just the opposite, in order to differentiate themselves. This leads to a patchy structure full of social cliques.

Figure 2.7: Patterns on animal skins.

The same type of interaction leads to the patterns that are often found on the fur of both predator and prey mammals: zebras, giraffes, tigers, and leopards (Figure 2.7). The striking patterns of spots or stripes of color on these animals are much larger than the size of a single biological cell—if the pattern were of the size of the cells then the animals would appear gray in color. It turns out that these patterns arise from interactions between individual biological cells that are very similar to the interactions between teenagers in high school.

We can think about how the cells of the skin might influence other cells to form patterns. The skin cells of these animals can emit chemicals into the fluid between them and influence the chemical activity of other cells. These interactions can lead to skin patterns. The chemicals affect not just the cells that are immediately adjacent, but all the cells in a larger area, whose size is determined by how fast the chemical moves (diffuses). There are two possible types of interactions, *activating* and *inhibiting*. When a cell producing pigment gives off a chemical that causes other cells also to produce pigment, we say that the interaction is activating. When a cell producing pigment gives off a chemical that causes others not to produce pigment, we say that it is inhibiting. An activating interaction causes cells

Figure 2.8: Formation of striped patterns through local activation and long-range inhibition.

to behave the same way, while an inhibiting interaction causes cells to behave the opposite way.[6]

Spotted or striped patterns can arise when there is a local interaction (which we can think of as caused by a slow moving chemical) that is activating, and a longer-range interaction (which is caused by a faster moving chemical) that is inhibiting. The activating influence causes cells that are nearby to form patches, like the children at circle time, or the panic in the auditorium. The inhibiting interaction tends to limit the size of the patches. Eventually, a stable pattern is reached. The dynamics of this model is shown in Figure 2.8.

Actually, this kind of interaction can lead to several different types of patterns. The differences arise from a bias that makes cells have a greater tendency to being dark or light. By changing the bias we can move from having light spots on a dark background to dark spots on a light background. When the preference for being light or dark is about the same, we end up with striped patterns. Figure 2.9 illustrates some of the final patterns that can result from different biases. These models are called "local-activation long-range-inhibition" models and they can help us understand many other kinds of natural patterns—domains in magnets, clouds, ocean waves, traffic jams, and even heartbeats.

The patterns on animal skins, formed during development, are very simple compared to the intricate structures of tissues and organs that are

Figure 2.9: From spots to stripes: some final patterns arising from local activation and long-range inhibition with different biases.

also formed during the same period. Still, the simple process of forming animal skin patterns captures an important feature of biological pattern formation: differentiation—parts of the system become different from each other, in this case different colors. In development a single type of cell leads to the creation of an organized arrangement of different types of cells. This used to seem mysterious. However, these examples show that even very simple rules of interaction can cause differentiation and can easily lead to remarkable patterns.

Chapter 3

Networks and Collective Memory

Patterns in networks[7]

Patterns in the brain are similar in many ways to patterns of people interacting in society. The patterns in the brain arise from interactions between neurons—the cells that comprise much of the brain. This is just as in society, where patterns of action are based upon interactions between people.

We use patterns in our brain to recognize and make sense of patterns in the world. The light that hits the retina of the human eye causes patterns to form in the neurons of the brain; and the relationship between the patterns in the brain and patterns in the world helps us to find our way around. We act when patterns in the brain are related to patterns of motion, such as the movement of arms and legs.

The patterns that form in the brain are different from the simple patterns we discussed in the previous chapter. The elements of the brain—neurons—are not just connected to nearby neighbors but are also connected to neurons that are farther away (Figure 3.1). We call this more complicated arrangement of connections a network

Neurons in the brain have many diverse forms and behaviors. For our purposes their behavior can be simplified to just two states: active ("firing") and quiet ("quiescent"). Neurons affect each other through connections called synapses. Synapses can be either excitatory or inhibitory. These are like the activating or inhibiting interactions of color cells discussed

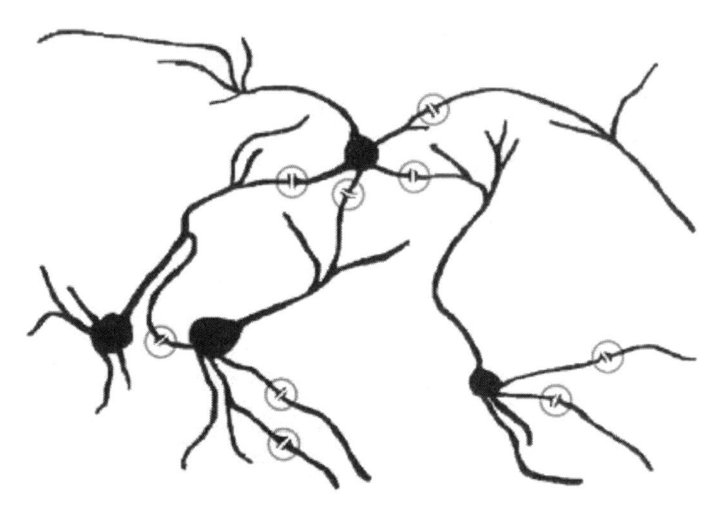

Figure 3.1: Neurons and synapses.

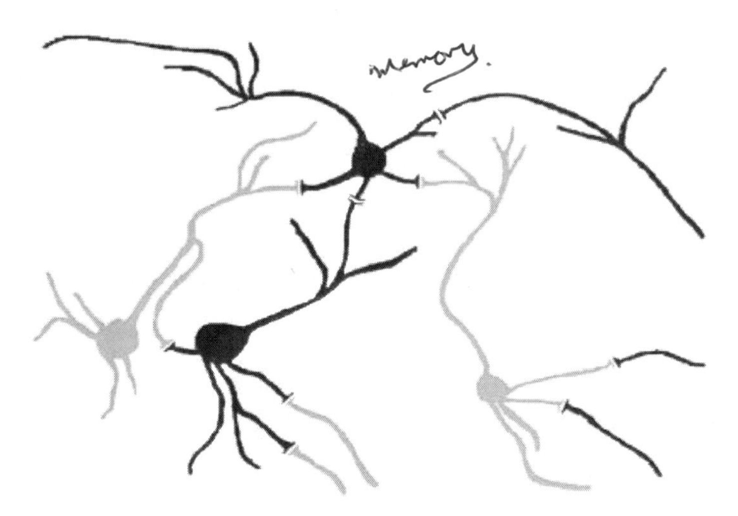

Figure 3.2: The pattern of neuron activity at a given moment. Black is firing; gray is quiescent.

earlier, or the influences between people that cause them to do the same or opposite of what their neighbor is doing.

An active neuron makes it more likely that another neuron will become active through an excitatory synapse (like an activating influence). On the other hand, an active neuron makes it *less* likely that another neuron will become active through an inhibitory synapse (like an inhibiting influence). In this way, a neural network is quite similar to the models of animal skin patterns or social influences discussed in the last section. The main difference between the neural networks and the models we have discussed in the last section is that synapses can connect cells that are far apart, moreover the excitatory and inhibitory synapses are not arranged in as straightforward a manner as they are in the local-activation long-range-inhibition model. Nevertheless, we can still discuss the pattern of firing of the neurons at any instant (Figure 3.2) like the pattern of pigment observed at an instant. Think of the "state of the mind" at a particular time as the activities of all the neurons—the pattern of neural activity. Imagine the pattern of lights that are on or off in a city at night. If you could see into your brain, this is what the activity pattern of neurons would look like. The pattern of firing of the neurons changes over time, just as the animal skin patterns changed over time, due to the influences that neurons have on each other.

The pattern of firing of neurons also is related to what is in the world around us and to our actions. The pattern of firing of neurons is influenced by the external world through the activity of sensory neurons that are affected by sensory receptors. These include the five usual senses—sight, hearing, touch, taste, and smell. A person's actions are effected by the influence of motor-neuron activity on the muscle cells. This means that the activity of certain neurons is related to the actions of the muscles. Thus, if we specify the activity pattern of the neurons, we are also (in large part) specifying the behavior of the person.

The synapses through which neurons affect each other are in part "hardwired" when we are born, but memory and experience also change them; this is how we learn. The simplest kind of adaptive learning is called *Hebbian imprinting*. Simply put, when two neurons are both active at a particular time, an excitatory synapse between them is strengthened and an inhibitory synapse is weakened. The same would happen if both were not active. However, when one is active and the other is not, the inhibitory synapse is strengthened and the excitatory synapse is weakened. Intuitively, what's going on is that the synapses become more "consistent" with the pattern through habit. Once the synapses have been modified to reinforce the pattern, it becomes possible to reconstruct the neural activity pattern from just a part of it. The imprinted pattern of neural activity becomes a memory.

To understand how this works as a memory, imagine that a picture is imprinted on the network of neurons. Then part of, or a modified version of this picture is shown to the network. The network will, through the influence between neurons, reconstruct/retrieve the original imprinted

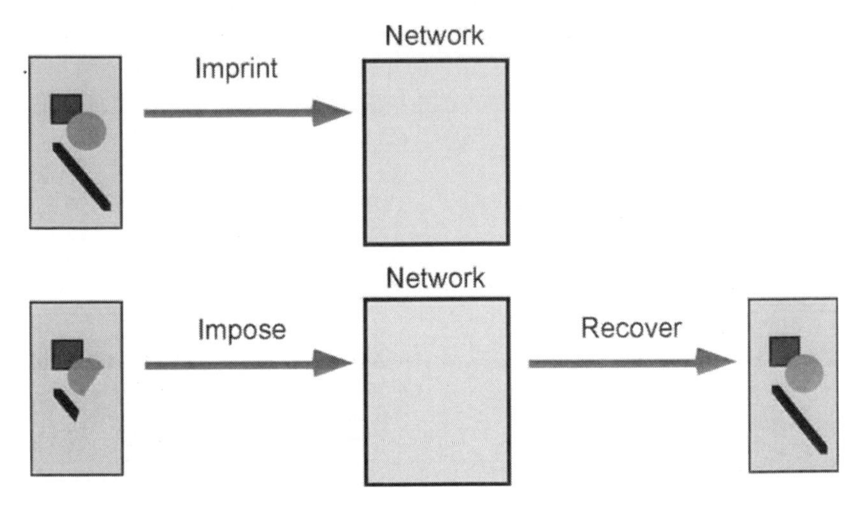

Figure 3.3: Recalling a pattern through Hebbian imprinting. Once the pattern has been imprinted on the network, it can "remember" the rest of the picture if presented with an incomplete or altered version of it.

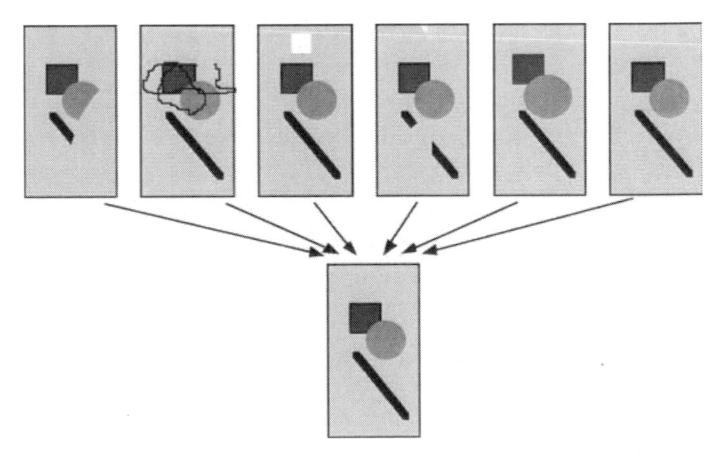

Figure 3.4: Some of the different patterns, similar but slightly altered from the original, that will trigger the same memory.

picture and thus "remember" the rest of the picture (Figure 3.3). It doesn't matter which part of the picture is shown to the network, as long as enough of it is shown, the rest will be recovered.

This neural network memory is called an associative or a content addressable memory. The imprinted state is retrieved using part of itself. Recovering the original pattern "associates" the reconstructed part of the pattern with the part of the pattern that was imposed. There are many patterns—similar to the original pattern but with slight differences—that will trigger the same memory (Figure 3.4).

This is one of the ways advertisers get you to think about their products. By imprinting images of their products on your brain strongly and repeatedly, they cause you to remember it when you see something similar (or anything having to do with their advertisement).

Network memory has very different properties than computer memory. In a computer the memory is accessed by an address that specifies the location of a particular piece of information. In order to retrieve information it is necessary to have the address, or to search systematically through the possible locations. If you ask a human being to quote line 64 from act 3, scene 1 of William Shakespeare's *Hamlet*, it's not likely that he will be able to come up with the right line. However, tell the same person to complete the line "To be, or not to be..." and he'll probably be able to recall at least the next few words immediately. A computer, on the other hand, will have a much easier time recalling the line by the former method, because it stores information according to its address. The discovery that network memories work in a similar way to human memory has led to a lot of excitement about using networks to model human thought.

The capacity of a memory is one of its most important properties. The set of patterns that will result in recall of a particular imprinted pattern (called the "basin of attraction") takes up part of the space of all possible patterns. If we try to imprint more than one pattern on the network, there is a limit to how many patterns we can imprint before the basins of attraction will interfere destructively with each other. When the destructive interference is complete, the basins of attraction disappear, and memory is not possible at all. The network capacity grows with the number of connections (synapses) in the network. If all of the neurons are connected to each other, then taking a network that is twice as large will lead to many more connections—enough so that the network can store twice as many independent images.

Using the network concept and its relationship to memory and learning we can consider other aspects of the way people think and develop a basic understanding of how the properties of our minds emerge from the properties of our brains.

Subdivision and creativity[8]

It has become generally accepted that one of the basic differences between human beings and computers is that human beings are capable of creativ-

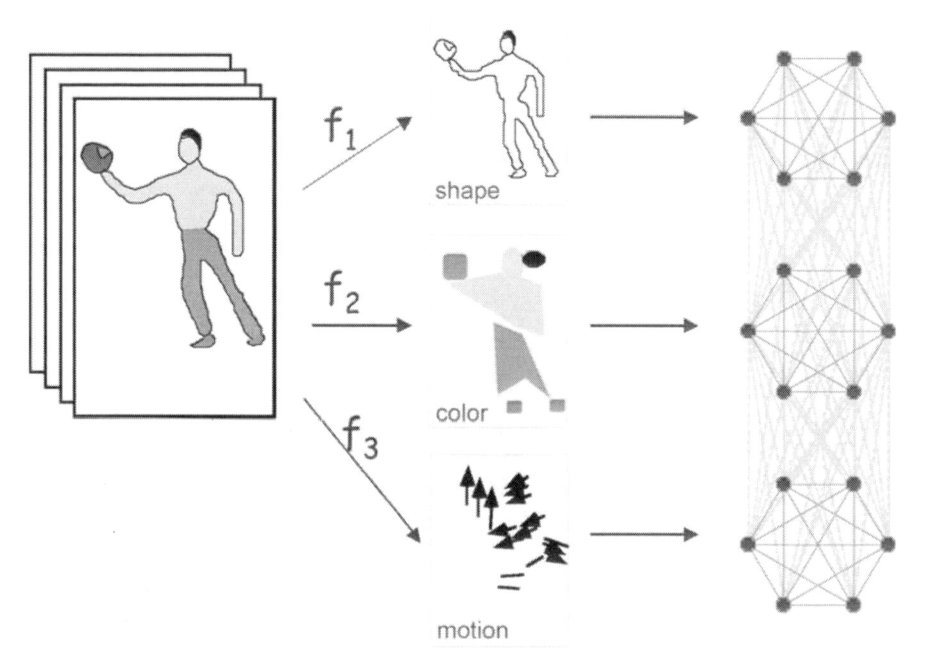

Figure 3.5: The separation of visual information into shape, color and motion. The information is stored in three separate subdivisions of the visual cortex.

ity. However, creativity itself has been somewhat of a mystery. How are human beings able to conceive of things they've never seen before?

The answer lies in one of the most interesting features of the brain: the way it's subdivided into different regions. Each of the subdivisions has particular functions, for example there are areas related to visual, auditory, and motor activity. If the brain was a fully connected network, every neuron would be connected to every other neuron, but the brain's subdivisions have far fewer connections between them. Why would this be the case?

People generally assume that networks with more connections are more powerful. It is true that the storage capacity of the neural network depends on the existence of connections; and if we reduce the number of connections, the ability to store patterns decreases. If that is the case, why is the brain subdivided at all? The reason is that when aspects of the world around us are partially independent, it is much better to store and act on them using partially independent parts of the brain. This is an important part of understanding how systems should be organized. It demonstrates why and how the function of the system dictates its structure. In particular, it explains the essential role of substructure, subdivision and specialization by functional segregation in subdivisions. To examine this concept more carefully, let's consider two examples of memory in the brain.

We will consider first the visual cortex, the main part of the brain related to vision. The visual cortex is separated into three parallel channels which, roughly speaking, specialize in information about color, shape and motion.

The reason for the separation is that color, shape and motion are partially independent features of the world around us. Different shapes can have the same colors, and the same shapes can have different colors. Objects moving in a certain way can have many different colors and shapes.

Because of this independence, it makes sense to describe objects using three attributes—color, shape and motion. There are many possibilities for each of them:

Color: RED, GREEN, BLUE, ORANGE, PURPLE, WHITE, BLACK, ...

Shape: ROUND, OVAL, SQUARE, FLAT, TALL, ...

Motion: STATIONARY, MOVING-LEFT, MOVING-RIGHT, RISING, FALLING, GROWING, SHRINKING, ...

Directing color information to one subnetwork, shape information to the second, and movement information to the third, lets us use *composite* patterns to identify objects: RED ROUND MOVING-LEFT, RED ROUND FALLING, BLUE SQUARE MOVING-LEFT, or BLUE ROUND FALLING. The pattern of neural activity in the color network identifies the color, the pattern of neural activity in the shape network identifies the shape, and the pattern of neural activity in the motion network identifies the motion.

Shape, color and motion are not entirely independent, however. Tree trunks don't move the same way or have the same color as leaves on the tree. Synapses that connect neurons in the different parts of the brain allow us to learn that certain shapes move in certain ways, or have certain colors. Having some connections between the subdivisions is helpful as long as there are not so many that composites are prevented from taking place.

Subdivision also helps our brains store and process language. Consider storing simple sentences in two kinds of networks. In the first case we store the sentences in a fully connected network and in the second in a network subdivided into three parts.

The number of short sentences that can be stored in the fully connected network (nine in Figure 3.6) is three times the number of patterns that can be stored in the subdivided network (three in the figure). We can only store three because we've broken up a network that could store nine into three separate parts. Each of the three parts stores one-third as many

Fully Connected Network
Imprint and recall

Big	Bob	ran.
Kind	John	ate.
Tall	Susan	fell.
Bad	Sam	sat.
Sad	Pat	went.
Small	Tom	jumped.
Happy	Nate	gave.
Mad	Dave	took.
Tired	John	slept

Subdivided Network
Imprint and recall

Big	Bob	ran.
Kind	John	ate.
Tall	Susan	fell.

Big	Bob	ran.
Big	Bob	ate.
Big	Bob	fell.
Big	John	ran.
Big	John	ate.
Big	John	fell.
Big	Susan	ran.
Big	Susan	ate.
Big	Susan	fell.
Kind	Bob	ran.
Kind	Bob	ate.
Kind	Bob	fell.
Kind	John	ran.
Kind	John	ate.
Kind	John	fell.
Kind	Susan	ran.
Kind	Susan	ate.
Kind	Susan	fell.
Tall	Bob	ran.
Tall	Bob	ate.
Tall	Bob	fell.
Tall	John	ran.
Tall	John	ate.
Tall	John	fell.
Tall	Susan	ran.
Tall	Susan	ate.
Tall	Susan	fell.

Figure 3.6: Content versus Grammar: Storing sentences in a fully connected network versus storing them in a subdivided network. The first can store more full sentences than the second, but the second can recall many more sentences than the three that were originally imprinted through composites.

patterns, or just three words. However, the divided network as a whole can now remember twenty-seven composite sentences. This is because every possible combination of three words stored in each of the separate networks is like a memory. The fully connected network remembers the specific sentences that are stored, but the second network recognizes all grammatically correct sentences made from these words. Learning only three sentences was enough to learn the many other grammatically correct possibilities.

The actual process in the human brain lies somewhere between these two extremes. Sentences make sense or are "grammatically correct" if properly put together out of interchangeable parts—words. However, a specific combination of words is used to describe a recalled event. Our brains do have some connections between subdivisions that store different parts of speech.

Notice that we stored three complete sentences (nine words total) in the subdivided network and ended up with a network that could remember twenty-four sentences that it had *never* seen before. Understanding subdivided networks can bring new insights into how creativity works. Consider a person who sees a bird flying and later sees a man walking. The shape of the man and the shape of the bird are stored in one part of the brain; the movements of the man and of the bird are stored in a different part of the brain. As a result, this person's brain can now imagine a composite pattern of the bird and the man: a flying man. This is the essence of creativity: creating new possibilities out of combinations of what already exists. When we speak or write we create new sentences out of words that we learned in other sentences. We take this kind of creativity for granted, but all creativity is a process of putting old parts together in new ways. The same notion of creativity applies in many parts of the natural world. Sexual reproduction, for example, creates new combinations of genes that can lead to formation of new biological organisms.

There are some important insights here about the organization of complex systems that we'll draw upon later. Most people don't understand the trade-off between independence and interdependence. When the parts of a system are independent, those parts are free to respond to independent demands of the environment. However, when the demands on one part of a system are linked to the demands on the other part, those parts will only perform well if they are connected to each other.

This discussion also helps us understand why social systems need to be organized in certain ways in order to be effective at their tasks. Independence between certain groups is important because it frees each of them to respond to independent demands of the environment. Only when the demands on one group are linked to the demands on the other group should they be connected to each other. This means that only when collective behavior is necessary should the parts be connected to each other and not otherwise.

Chapter 4

Possibilities

Introduction[9]

Imagine a flower, a chair and a person. Imagine how you would describe each of them. If words fail you, consider a photograph or a movie.[10] Words, a photograph, or a movie can all be used to answer the question "What does it/he/she look like?" Descriptions underlie everything from science to art. Science explores the descriptions we share (or should share) when we look at the world. Art explores the differences between the descriptions that exist in each of our minds. Thinking is always about descriptions even when we don't realize it because what we have in our minds is a kind of description, not the system itself.

Even when we have a simple pattern, like animal skins, it is hard to know exactly how to describe it in words. Saying that we have spots or stripes helps, but what about the details of their locations? What about the details of the shapes? A complex system is hard to describe and the ability to describe it is central to our ability to understand it. Imagine that we have to learn about a system by studying a description of the system. The longer the description, the longer we would have to study it. This makes it natural to define the complexity of an object as the length of the description. An object that is more complex has a longer description, while a simpler object has a shorter description.

The notion that complexity is measured by the length of the description seems, however, to suggest that complexity is a very slippery thing. If I am describing something to another person, the length of a description that I need depends on what the other person knows, and even what language we are using. The idea that complexity is not an absolute, but is relative to who is giving the description and who is receiving the description should not discourage us from thinking about complexity. Descriptions are always relative to the observer and this is already recognized in basic physics.

For example, the speed at which something is moving is relative to the observer. If you are going in a car at 60 miles per hour and the car that is next to you is also going in the same direction at 60 miles per hour an hour according to its speedometer, it doesn't seem to be moving at all, as far as you are concerned. On the other hand, if you were moving in the opposite direction at 60 miles per hour the same car would appear as though it is going 120 miles per hour. One of the main ideas of Mechanics (the study of objects in motion based on Newton's Laws) is that we can relate what one moving observer sees to what another observer sees, even when what they see is different because they are not moving at the same speed.

The idea of relating what different observers see was made into a principle by Einstein in his theories of relativity. He thought about observers who were not only moving at different speeds (the subject of special relativity) but also speeding up or slowing down. Accelerating upwards (like in an elevator or in a rocket) makes a person feel like the gravity is stronger. This relationship between accelerating observers and gravity is the basic idea behind general relativity.

If the degree of complexity is relative, our job is to describe how different observers measure complexity. Each observer considers the complexity of a system to be the length of a description that he needs to describe the system. Because of differences in the observers the length of their descriptions differ. The trick is to understand the systematic way the lengths differ so that this variation can be part of our understanding. In this section we focus on what happens when observers use different languages. Then in the next section, we will focus on the level of detail we choose to provide in the description.

Fifty years ago, Claude Shannon, a mathematician at Bell Labs, discussed the problem of communication in a way that still forms the basis of our understanding today. He answered the question of how long messages in different languages have to be to say the same thing.[11] Shannon recognized that messages in one language are longer or shorter than messages in a second language in a specific way that can be determined by counting the number of possible messages of a certain length. The idea of thinking about all of the possible messages (the space of possibilities), instead of just a specific one, is a key idea. If you have a message in one language, say English, and you want to translate it into another language, say Japanese, how long will the new message be? First, you must determine how many sentences in English have the same length as your message. Then, you must figure out how long sentences in Japanese have to be to obtain the same number of possibilities. This is the length that the translated message should be. Does this seem like a round about way to figure out how long the translation will be? Of course, for one case it is round about, but it answers the question once and for all for all possible messages. Shannon's discussion of possibilities (the space of possibilities) is helpful in understanding many issues. Here, we will apply it to thinking about complexity.

Consider the problem of describing something to a friend. In front of you is an object. If you want to describe this object, you have to identify (pick) it out of the many possible objects that could be in front of you. In order to be able to identify this one out of all the possible objects, the number of possible descriptions has to equal the number of possible objects. Then each of the possible objects can correspond to one of the possible descriptions.

Figure 4.1: An object S and its description. The length of the description is closely related to the complexity of the object.

Let's say that there are M possible objects, how long does the description have to be so that we will have enough possibilities? The length of a description is related to the number of possibilities. The longer the message, the greater is the number of possibilities that exist. Today, we often think of storing information in computers. Computers store information in "bits." Bits are like light switches that can be on or off. Each bit has two possibilities. Two bits have four possibilities. Three bits have eight possibilities. Four bits have sixteen possibilities. Every bit we add increases the number of possibilities to twice as many as before. Multiplying rather than adding means that the number of possibilities grows very rapidly. For example, one hundred bits gives about 1,000,000,000,000,000,000,000,000,000,000 possibilities.

What happens when we use sentences in English to describe something? It turns out that if we only count the sentences that make sense, then the number of possible sentences increases by roughly a factor of two for each additional English character. One character in English could be any one of the 26 letters, they could be capital or small, and there also could be punctuation. You might think that because there are many more letters than just two there would be more sentences than this. However, using real words, grammar and generally making sense, significantly limits the number of possibilities. Still, there are a lot of possibilities. Writing 100 English characters gives the same huge number of possibilities as 100 bits. Thinking about how many possible books there are is mind boggling, but amazingly, this is the kind of complexity that people's minds can absorb.

So, if you want to characterize the complexity of an object, think about how much you would have to write in order to describe it. Would it take a sentence, a paragraph, a few pages, a book, or many books? Count the number of characters in the description. This is its complexity.

Complexity and scale[12]

Describing systems in the world involves a decision about the level of detail we provide. The length of a description depends on how much detail we can see. If we are far away from an object, we can't see many details. The description would then be much shorter than if we were close to the object. Think of using a zoom lens to take a picture. If we zoom in on a person we see a lot more detail than if we don't. If we are far enough away, the person looks just like a speck (Figure 4.2).

The dependence of the complexity on the amount of detail is important enough that we will discuss several different cases, shown by the three

Figure 4.2: Description at different scales: three views of a person.

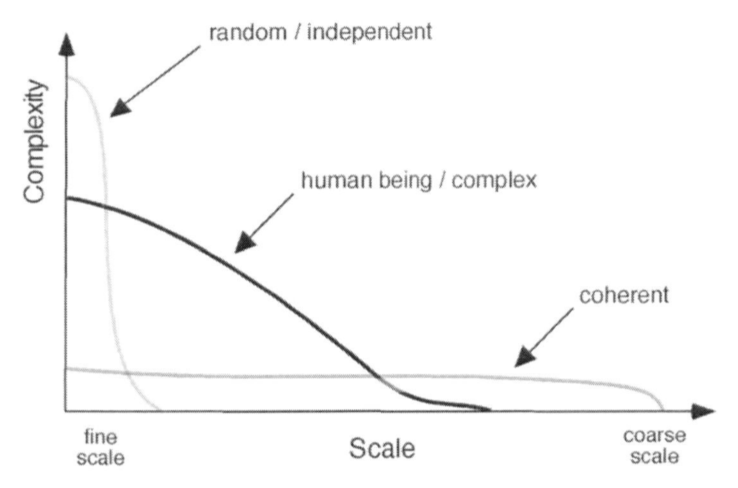

Figure 4.3: Complexity as a function of scale for three kinds of system: random, coherent and complex. The way a system is organized affects how it is seen at different scales.

curves in Figure 4.3. The horizontal axis indicates how far away you are from the object you are describing. Better yet, it indicates the level of precision (scale) of the description. The vertical axis indicates the complexity of what you are describing (the amount of information needed to describe it).

The curve marked "human being/complex" shows what would happen if we described a person. The closer we get, the more detail there is, and therefore, the longer the description. It is better if we think not only about describing a person at a moment, but rather about describing a movie of the person over, say, a day. Also we need to be able to ignore (see right through) the things that are around the person that might block our view.

When we are far away from the person, we would see only a point moving around. We might see the person go from home to work and back, go out to dinner or to the movies, or see the person go on a trip by airplane, but not much else. This would be interesting to a sociologist thinking about how people travel from place to place.

If we are closer, we would see all of this, but we could also see the person's legs and arms moving, walking around the room, going from room to room at home, or walking between places at work.

Still closer, we would see the person's mouth moving, hear what he/she is saying, see his/her facial expression, what his/her fingers are doing. This is the level of detail that we would normally see if we were standing at a distance natural for talking with the person.

For the purpose of considering complexity, we don't have to limit ourselves to this distance, we can consider much closer distances that are not generally practical. We also don't have to limit ourselves to using a regular camera, we can think of using a magnifying glass, or even a microscope.

Usually, when we think of a magnifying glass or a microscope, we look at only a small thing. However, for thinking about the complexity of a person, even though we are making a movie of the person with a magnifying glass, we still want the whole person in the movie. It would take a really large screen to do this. With a magnifying glass, we can, for example, see all of the pores and hairs on a person's skin. When we describe the person with this level of detail, we have to describe each and every one of the pores and hairs. The idea is not to describe each pore and hair separately, but rather to have a picture of the person that is large enough to see these very fine details while at the same time keeping the whole person in view. Of course this description would be very long.

It is even better to think about this as if it were a CAT scan, where we can see all of the internal parts of the person and what these parts are doing. Depending on whether we are looking with a magnifying glass or a microscope (how much magnification we use), we can see all of the organs of the body, or all of the cells of the body, or all of the molecules, or even all of the atoms. By the time we are thinking about writing a description of all of the atoms, it would take a remarkably long time to write the description. From physics we actually know how long a description this would be. If we cut up the entire earth into little pieces the size of grains of sand and wrote one English character on each grain of sand, there would be barely enough characters to write this description.

This is clearly a very long description. However, while it is very long, it is still "finite." This means that even if we describe a person atom by atom, there is a limited amount of information that we need. The reason this is true originates in quantum physics, which tells us that each atom has some uncertainty built into it. So we only need to say where it is with certain accuracy, and that is sufficient.

The curve marked "random/independent" in Figure 4.3 shows a different case. This curve would be the case if we took a person and mixed up all the atoms so that they were no longer organized in any particular fashion. The atoms would also not be moving in a particular direction, but would move in any direction. Each of the parts is acting randomly. If we put these atoms into a large vat, it would look like murky water. This is what physicists call "*equilibrium*." Looking at it from far away there isn't much to describe because it doesn't go anywhere. Even if we look much closer at the liquid, it still appears boring. The reason for this is that the random placement of atoms makes all of the parts look the same—the parts are undifferentiated. This is true up until we reach the scale of describing what each of the atoms is doing. What is special about this case is that all the atoms are moving independently. So when we want to describe what all the atoms are doing, we actually have a longer description than the one for the person. The equilibrium liquid is "more complex" than a person when we describe it at the level where we can distinguish the individual atoms. However, this is only true if we are describing the system at this level of detail. Otherwise the person is much more complex. The curve marked "random/independent" is higher than the curve marked "human being/complex" for very small scales, but otherwise it is lower.

The third case (marked "coherent" in Figure 4.3) is what would happen if we took the same atoms and organized them so that they were all moving in the same direction. It may surprise you to know that if your atoms were all moving in the same direction, you would move at a speed of about 2,000 miles per hour. The reason we don't move that fast is that the atoms are constantly bouncing against each other and are tied to each other by various kinds of chemical bonding. Of course, if we did organize them to move in the same direction, the motion would be visible from far away! This case we can call *coherent motion*.

The three cases—random, coherent and what we normally think of as complex—illustrate how the way a system is organized affects how it is seen at different scales. Visibility at a large scale means that things are organized. In order for us to see behavior at a large scale, the parts must be moving together. For example, we can see this in how a muscle works. Muscles have many cells doing the same thing at the same time. Because of the actions of muscles we perform motions that are visible at a large scale compared to the size of the individual cells. A human being has various groups of cells organized to work together. The groups are of many different sizes. Depending on the size of the group of cells working together, we see what they are doing on a different scale. This is why there is more and more to see as we get closer to a human being.

Thinking about random, coherent and complex systems applies to any kind of system, physical, biological or social. For example, a liquid in a cup is a physical system where atoms are moving randomly, a cannonball has atoms moving in an organized way, and the atoms of a snowflake are organized so that there is structure on many different scales. Among biological organisms, cells in a pond tend to move randomly, a bacterial infection involves many cells working together, and the cells of a human being are organized to have structure on many different scales. In social systems, people in a crowd move aimlessly, a mob or an army moves coherently, and a corporation has people organized to have structure on many different scales. If we think about the crowd of people moving in all directions, when one person moves one way, another person moves the other way. If we look from far away nothing seems to happen. In the case of a mob or an army we can see what is happening from very far away because the motions of the individuals add together. In the case of an organization, as we get closer we see more and more details about what is going on.

These examples all show a trade-off between large scale behavior and fine scale complexity. When parts are acting independently, the fine scale behavior is more complex. When they are working together, the fine scale complexity is much smaller, but the behavior is on a larger scale. This means that complexity is always a trade-off, more complex at a large scale means simpler at a fine scale. This trade-off is a basic conceptual tool that we need in order to understand complex systems.

In the next section we will devote more attention to the subject of social systems and how we can understand them using the properties of complexity and scale. Before we do this, let's consider again the complexity of a human being. This time let's think about how one person describes another. The person doing the describing is going to use his own senses (not a microscope) and is going to be located a distance away, of say a meter or two, which is how we usually interact with each other in social contexts. How much information would be necessary for this description?

We can estimate this by using the amount of memory needed to store a movie made by a regular video camera. These cameras are designed with people in mind, how sensitive our eyes are, and how sensitive our ears are to sound. It is easy today to take a digital video camera and plug it into a computer to see how much memory space a video would take. It turns out that about five minutes fills a gigabyte (a billion bytes) of memory, since a byte is about 10 bits, this is about 10 gigabits of memory, which is also about the space on a CD-ROM and about 10–20% of the size of a DVD. This suggests that a DVD can only store roughly 25 minutes of video. Actually, it can store about 2 hours of video by using compression. Compression eliminates the recording of parts of the picture that are not

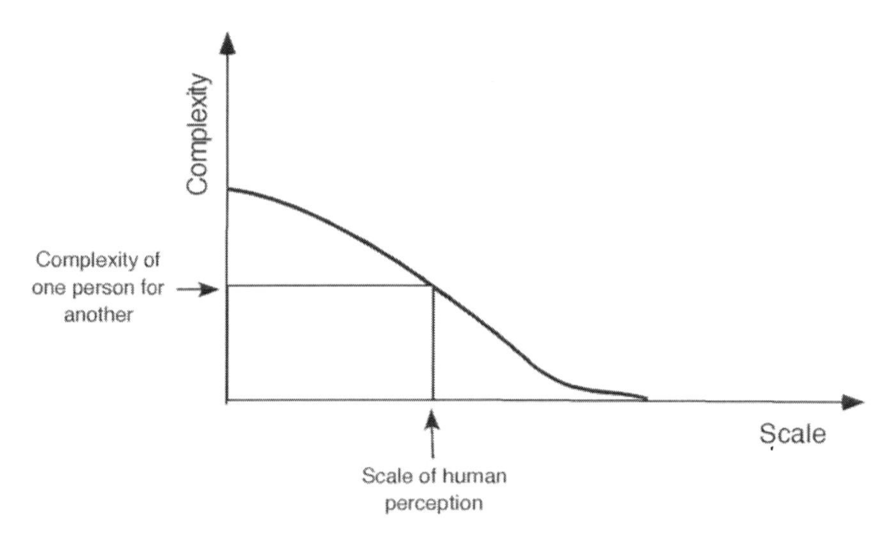

Figure 4.4: The complexity of a human being at the scale of human perception.

changing. If we extended a movie to a day, we would have enough video to fill about 10 DVDs or about 400 gigabits. To describe a person over a lifetime we would multiply this number by a typical number of days in a lifetime of 80 years, about 30,000. So it would take about 300,000 DVDs to store a description of a human being over the course of his/her lifetime. Of course, a person repeats many things that he/she does, so we could make a shorter description if we tried to. Nevertheless, this gives an idea of the complexity of a person from the point of view of another person.

While the specific value of the description length is not essential, the notion that the complexity of a human being is limited will be important when we discuss how complexity applies to social systems.

Chapter 5

Complexity and Scale in Organizations

Complexity of social systems[13]

The dependence of complexity on scale can be discussed for many different kinds of systems. Rather than thinking about the usual systems that are studied by the science of parts, it is particularly exciting to think about systems that conventional science doesn't have many tools to think about. Let's see what we can say about the most complex systems we know about: human organizations and human civilization as a complex system.

Why should we think about human civilization? Aside from the obvious, that we are all part of it, there is a specific reason to consider the complexity of human civilization: everyone seems to be complaining about how complex life is becoming.[14] This complexity is not due to any dramatic change in the natural environment. For example, trees haven't suddenly become harder to understand. What has become more complex is our social and economic systems. What can we say about the complexity of society?

To start thinking about this problem, we might notice that the world has become much more interdependent. This is what we mean when we refer to the "global economy." The interdependence means that something happening in one place in the world can, and often does, affect things happening in another place, even in many different places around the world. If things are more interdependent, the complexity of the world at larger scales has increased. Simply put, if we want to describe the world, we need to mention all of the things that have impact on a lot of people. Since there are many such things, there is a lot to describe.

Another approach to thinking about the complexity of society is to consider how the interdependence comes about. How do people influence each other? We think of influence between people as control, not necessarily coercive control, but control nevertheless. Traditionally, the way people influence/control each other is through organizational structure. This suggests that we consider how control works in companies, governments, and other social organizations. In traditional organizations, control is exercised in a specific way—through a hierarchy. For about 3000 years, hierarchies have been the generic form of human organizations. It is helpful for us to understand how a hierarchy works and what this means for the complexity of a social system.

To help us think about a hierarchy, it is useful to focus on an idealized hierarchy such as the one shown in Figure 5.1. In an ideal hierarchy, people only communicate up and down the hierarchy. If you want to do something with someone in the office next door to you, you must first talk to your boss and your boss will then tell your neighbor what to do. If your boss does not supervise your neighbor (or if you want to do something with the person down the hall), your boss must talk to his boss, who in turn will talk to the boss of the person in the office next door, who will tell him what to do. Of course, the bosses don't need to wait for someone in the ranks to suggest an action, they might just decide to tell a bunch of people what to do. Another way to think about the communication through the hierarchy is that the communication up the hierarchy filters the information that is needed for the bosses, while the communication down the hierarchy provides details that are needed for the workers.

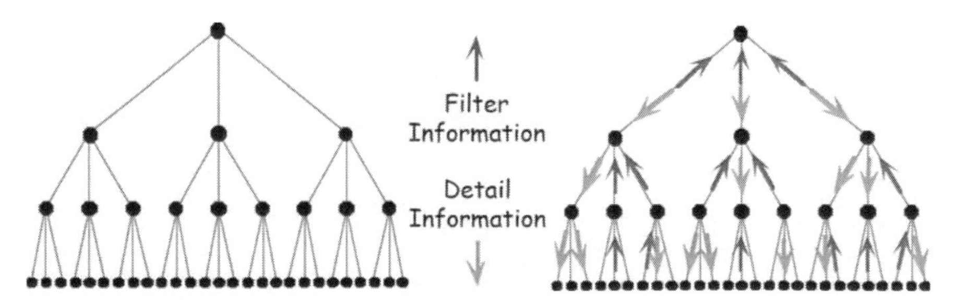

Figure 5.1: An idealized hierarchy. Information can travel only up and down the hierarchy.

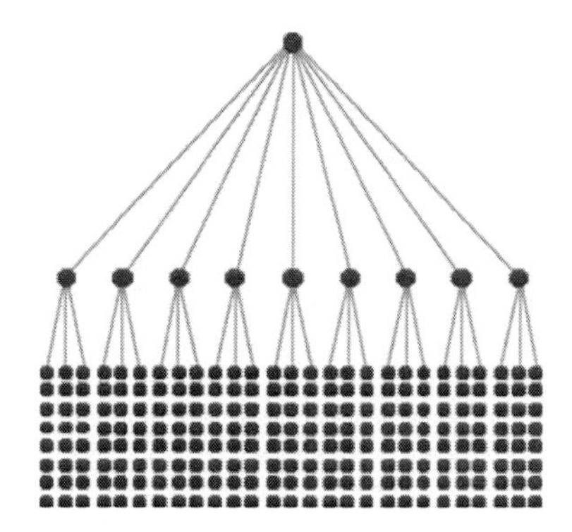

Figure 5.2: Another idealized hierarchy.

Hierarchies can differ from each other, particularly in how many individuals are supervised by a single boss (Figure 5.2).

To help us think about hierarchies, we need some examples of organizations and what they do. A couple of useful examples are a military force and an industrial factory. To illustrate the ideas, early versions, rather than modern versions, of each of these are particularly useful.

As an example of a military force, consider the ancient armies that conquered much of the ancient world, specifically Alexander the Great's Phalanxes or Roman Legions (Figure 5.3). These military forces are almost

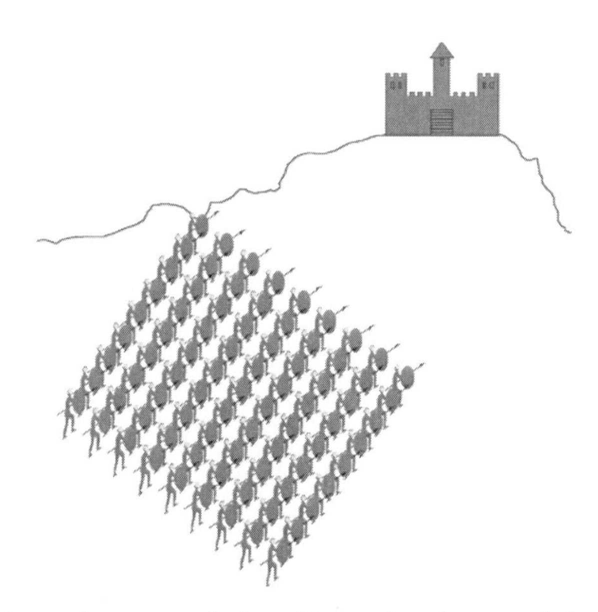

Figure 5.3: An ancient army, designed to produce large-scale impact.

like the idea of coherent motion that we discussed earlier in the book. Their behavior is characterized by long marches, with many individuals doing the same thing at one time, and repeating it many times. The behavior of each individual is very simplified. Here we can see the trade-off between complexity and scale. The construction of the Phalanx or the Legion was designed for large-scale impact. Indeed, the scale of impact of these forces was remarkable even by today's standards. Still, to be effective the military force has to respond to what is going on around it. For this there was a control hierarchy that determined the direction the military units should march. In the military hierarchy, many individuals can be under the supervision of a single commander.

For the next example, consider a factory, specifically a Model-T Ford factory (Figure 5.4). Ford's concept started from simplifying what each person was required to do. Each person performed a simple task and repeated it many times. Different people performed different tasks. These tasks were coordinated to produce a single product. The product could be quite complex, like a car, but the key idea was that the number of cars produced would be large—mass production. The scale of action of this system is large because of the repetition of simplified tasks. Again we see the trade-off between complexity and scale. In addition to the trade-off

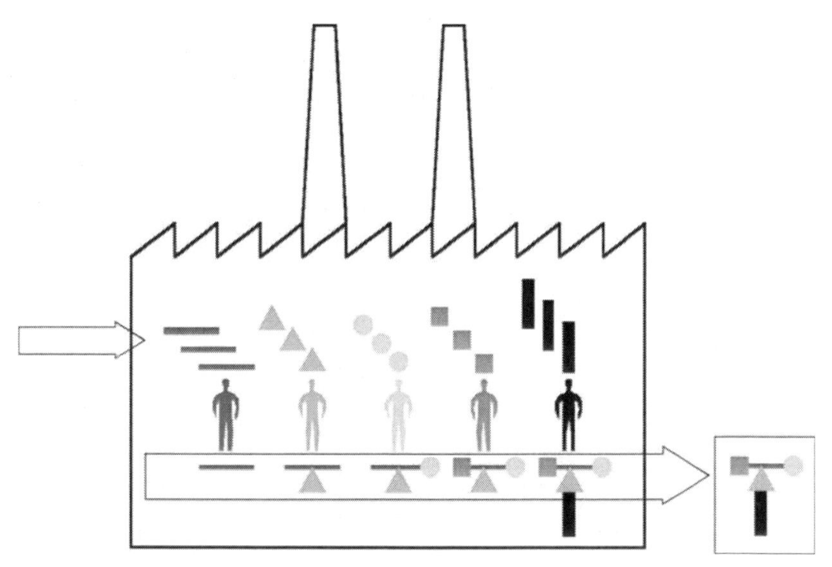

Figure 5.4: A schematic drawing of a Ford Model-T assembly line. Each individual performs a single, simplified task repeatedly, resulting in the large-scale production of cars.

between scale and complexity, we can also see the role played by the control hierarchy. The hierarchy coordinates the tasks of different individuals. Because individuals are doing different things, the control hierarchy has to give many more instructions than in the case of the military. Intuitively, this means that there must be fewer individuals directly supervised by a boss than in the ancient army.

Now that we have seen a couple of examples of hierarchies, let's consider the basic nature of the hierarchy itself. We can see that the hierarchy enables a single individual (the commander or CEO) to control large-scale behaviors. The CEO needs to know something about what individuals in the organization are doing. However, he/she does not need to know everything about what they are doing. Specifically, the CEO does not need to know every detail about what each person does every minute of every day. It *is* necessary for the CEO to know or to control matters that affect a large proportion of the organization. These are the organization's large-scale behaviors.

Another way to see this is to consider the communication through the hierarchy. Any communication that involves people in well-separated parts of the organization (the three groups circled in Figure 5.5) must go through the CEO or commander. This would be true of almost all large-scale behaviors.

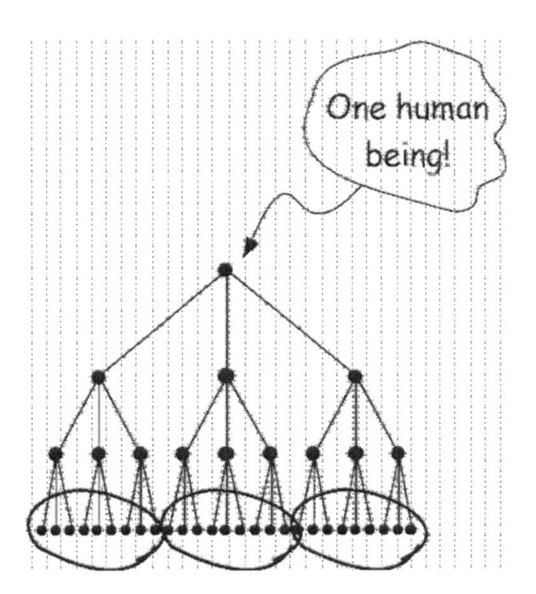

Figure 5.5: An idealized hierarchy. The collective behaviors of the organization are limited to the complexity of the individual at the top.

We've arrived at an important conclusion. Since the large-scale behaviors are communicated through the CEO, there is a limit to their complexity—the large-scale behaviors cannot be more complex than the CEO. This complexity is large, as large as a single human being, but it is limited. At most 10 DVDs of information are needed to describe what the CEO does in one day. This is a lot, but it is still a finite amount of information.

Let's compare the hierarchy with other types of organizational structures (Figure 5.6). One type of structure we can think about is a network, like the network of neurons in the brain. When we discussed the brain as a network, we did not think that one of the neurons was responsible for the large-scale behavior of the system. Each neuron could be simple and yet we could have very complex behavior of the network as a whole. We shouldn't think that any randomly connected network behaves in a complex way. Still, it is possible to have a network that together is more complex than its parts. This is not true of the hierarchy. We see that the hierarchy is good at amplifying, increasing the scale of behavior of an individual. However, it is not able to provide a system with a larger complexity than that of its parts.

Real organizations today are not pure hierarchies, they are hybrids of hierarchies and networks. There are many lateral connections corresponding to people talking to each other and deciding what to do. Nevertheless, from this discussion we can learn the following key point: to the extent that a single individual is in control of an organization, the organization is limited in complexity to the overall complexity of that single human being. Is this an important observation? To answer this we need to understand why an organization (or any other system) needs to be complex.

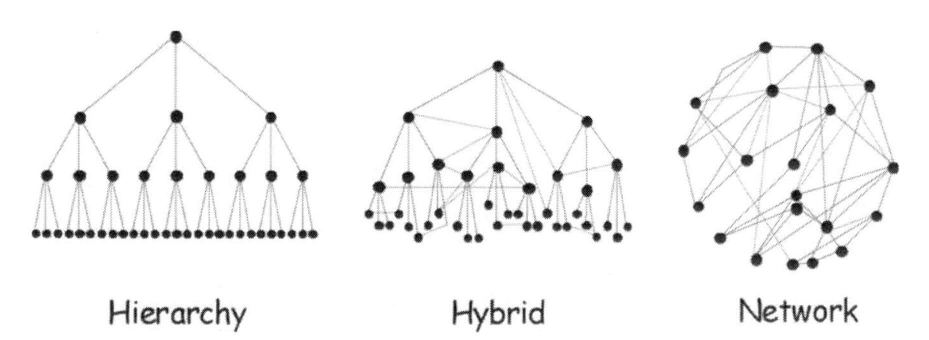

Figure 5.6: Three types of control structure.

Why complexity?[15]

Why is it helpful to be complex? The answer is that being complex is the only way to succeed in a complex environment. This of course begs the question: What is a complex environment? A complex environment is one that demands picking the right choice in order to succeed. If there are many possibilities that are wrong, and only a few that are right, we have to be able to choose the right ones in order to succeed. As a general rule this requires a high complexity.

Consider the viability of the offspring of different types of biological organisms. Most types of animals have many offspring. The number of offspring that survive to adulthood tells us something about how complex an animal's environment is compared to its own complexity. Mammals have several to dozens of offspring, frogs have thousands, fish have millions and insects can have as many as billions. In each case, on average only one offspring per parent survives to have offspring. The others made wrong choices because the number of possible right choices is small. In this way, we can see that mammals are almost as complex as their environments, while frogs are much less complex and insects and fish are still less complex when compared with their environments.

Although Darwin's theory of evolution discusses how the fitter offspring tend to survive, the reality is that whether or not an offspring will survive is mostly a function of chance due to the many possible wrong choices that exist for each right choice. Higher complexity organisms have more behavioral options, which in turn enables them to make more right choices.

Although complexity is very important for survival, scale also matters. In general, larger scale challenges should be met with larger scale responses. The rule of thumb is that the complexity of the organism has to match the complexity of the environment at all scales in order to increase the likelihood of survival.

The same argument can be made in the context of economic systems. If the environment of a corporation is very complex, many decisions must be made correctly in order to succeed. These decisions might include: product choices, price decisions, investment choices, resource allocations, hiring policies, mergers and acquisitions, and so on. Students of economics and management are taught how to make such choices in order to increase the likelihood that they will make the choices that lead to success. The best a single person can do, however, is limited by his/her complexity.

A key to the problem of corporate success is that companies are competing with each other. This means that if one company makes better choices than another, it will succeed and the other will tend to go out of business. Both scale and complexity matter, larger scale companies and more complex companies will tend to succeed. This leads to a kind of "arms race" where companies that increase their scale or complexity, tend to succeed at the expense of other companies.

The same ideas also apply to military power and the appearance of ancient empires. Why did one country take over another country to become an empire? The answer is generally because it had a larger scale or more complex military. We can combine thinking about scale and complexity in all of these examples by using the curves we discussed earlier, which show complexity as a function of scale.

Consider the two pictures in Figure 5.7. Comparing these pictures illustrates the combined issues of complexity and scale in two successful organisms. The legs of a wolf are designed for the largest scale action: moving the animal as a whole. The structure of a person gives up some of the ability to run fast. Only two of the four limbs are for moving the entire organism. The arms and hands are designed for finer scale, higher complexity, manipulations. If the environment requires larger scale motion/action the wolf is better suited for that environment. If the environment requires a finer scale higher complexity manipulation, the person is

Figure 5.7: Limbs and locomotion: Trade-offs and effectiveness in scale and complexity.

better suited. This figure therefore demonstrates two key ideas: 1) there is a trade-off between complexity and scale, and 2) the success of the organism/organization depends on both complexity and scale.

Chapter 6

Evolution

Selection and competition

The self-organization of patterns that we discussed in Chapter 2 can explain the formation of spots and stripes and a variety of simple patterns. The human form is established during fetal development through many successive layers of such simple patterning processes. However, in order to understand how the layers have been combined together so effectively to make a complex organism, we cannot just think about pattern formation itself. We also need to think about what caused that particular combination of patterns to occur. The theory of evolution provides an explanation of how patterns can be determined and combined together to form complex organisms through changes that occur over many generations. Previously, we saw that human organizations (governments, corporations and other social organizations) may be thought of as undergoing a kind of evolutionary change through survival of the most effective organizations. To understand better how this works, we can study the general principles that underlie biological evolution.

Biological evolution is a process by which populations of organisms are transformed over time. Organisms do not change individually, rather the changes occur between one generation and the next. Different individuals have different traits. Some of the individuals reproduce more than others, and some end up surviving to maturity more than others. Because traits are hereditary, the relative rates of reproduction cause the faster-reproducing and better-surviving types to dominate the slower-reproducing and poorer-surviving types over many generations. This process is called natural selection.

Since Darwin first proposed the idea of evolution by selection,[16] biologists have focused on competition as the driving force of evolutionary change. Conceptually, the competition is between organisms for rapid reproduction and survival to maturity. A limitation on the number of organisms that can exist at one time also plays an important role. The availability of resources and the way that the organisms interact with each other limits how many of them can coexist in the same generation. The ones that are successful in having offspring are the ones that are better at surviving at that time. This comparison of organism survival is interpreted as a competition. Because of this, the idea that cooperation occurs among animals has seemed antithetical to evolutionary ideas for some time.

In order for the process of Darwinian evolution to work, there are several conditions that must be met. Offspring have to be similar to their parents through heredity. Still, in order for changes to continue over many generations there also have to be some differences between offspring to allow for ongoing changes, i.e., some variation. These two properties are manifest and were well known to Darwin in the artificial breeding of animals and plants. The mechanisms by which they occurred, however, was not known until after Darwin's time. Today we know the mechanisms center around the inheritance of DNA molecules, as the sequence of DNA carries much of the information from generation to generation leading to heredity. Changes in this sequence lead to opportunity for variation. DNA sequences are organized into long strands. Often we consider the smallest functional part of a strand to be a gene. A gene contains the information needed to make a particular protein. Once the protein is made, it acts like a molecular machine, an enzyme, which is the basic building block of the mechanisms of cellular functioning.

The picture in Figure 6.1 and the ideas behind it were, however, not entirely satisfying to biologists a hundred years ago because many biological organisms, both animals and plants, reproduce sexually. This complicates the picture because reproduction creates organisms that can be quite different from their parents. Offspring receive some of their genes from each of their parents and their traits are a kind of composite of the traits of their parents. Figuring out how this relates to Darwinian evolution is not a simple task. To solve this problem, Ronald Fisher, John Haldane

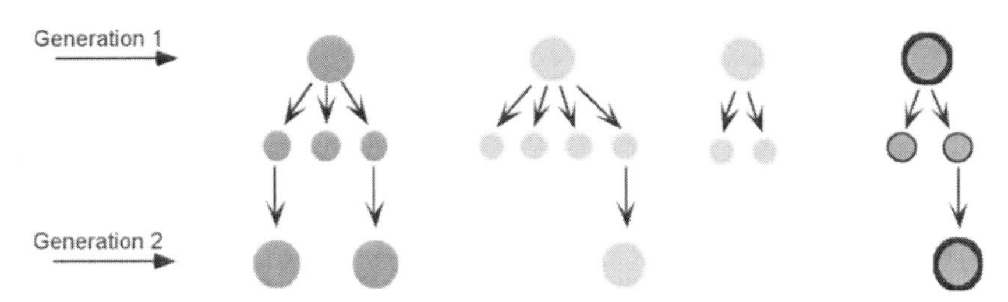

Figure 6.1: Natural selection from generation to generation.

and Sewall Wright[17] formulated the Neo-Darwinian picture, the dominant theoretical framework in which evolution has been discussed since the early part of the last century.

You may be familiar with a book by the British biologist Richard Dawkins called *The Selfish Gene.*[18] This book popularized the basic concepts of Neo-Darwinism. According to this perspective, evolution should be described as a competition between genes—and that's that. Any forms of cooperation between individuals or groups of individuals arise merely through the hidden agendas of these Machiavellian genes.

Neo-Darwinism helps solve the problem of thinking about sexual reproduction because genes themselves don't reproduce sexually, they replicate. If we think about evolution as a competition between genes, so that the circles in Figure 6.1 are the genes inside an organism, then sexual reproduction of the organisms doesn't affect the basic process shown in the figure.

The idea of competition between genes is not only based on general ideas but on a specific mathematical argument that was developed by Fisher and explained quite simply by Dawkins. He explains it using a "rowers analogy."

In the rowers analogy, we think about races between teams of rowers in boats. Each of the rowers is analogous to a gene, and each of the boats is analogous to an organism. Dawkins used this analogy to explain why we might think of a competition between rowers instead of a competition between boats. Basically, the idea is that even though the boats are competing with each other in a single race, if we run many races, we end up comparing rowers. There are several examples that Dawkins uses to explain particular features of this kind of competition. Let's examine one of them more closely, the competition between English speaking and German speaking rowers.

Dawkins describes a set of rowers, a "rower pool," that are placed into boats in sets, with all of the boats having the same number of rowers. The boats run heats against each other and the winners are placed back into the rower pool to compete again. Adding a little from the evolutionary picture, we imagine that the rower pool remains about the same size because the rowers "replicate." After each race, the successful rowers are replicated enough to replenish the rower pool. This step makes up for the rowers that lost, so that we always have the same number of rowers in the rower pool and in each heat. We further imagine that there are two kinds of rowers, English speaking and German speaking. The languages play a role in how well a boat does because we assume that when mixed boats compete against one-language boats, the one-language boats have an advantage (because the rowers can understand each other) and win.

What will happen over time to the rower pool? If there are more English speaking rowers, there is a higher probability that a boat will have all English rowers. Moreover, German speaking rowers will tend to have English speaking partners. This means that English speaking rowers will win the race more often than German speaking rowers. Over time, the number of English speaking rowers will grow and the number of German speaking rowers will shrink. Eventually there will be an all English speaking rower pool. Alternatively, if we start out with a rower pool that has more German speaking rowers, over time the number of German speaking rowers will grow, and we will end up with an all German speaking rower pool. In either case, we can think about this as a competition between the rowers, with one type of rower winning over the other type over time.

Dawkins' argument seems quite reasonable, but there are hidden assumptions in this argument that are important to discuss. We will analyze this example and use it to consider whether, or better yet under what circumstances, the Neo-Darwinian view, that gene competition is enough to describe evolutionary process, is correct, and under what circumstances it is wrong.[19] This does not mean that the Neo-Darwinian view is not a useful and powerful way to think about evolution. However, finding out that it is not always correct means that selfish genes are not all that there is to evolution. This is an important result precisely because the competition of genes (and the idea that competition is what is important) has been used to motivate many ideas about biology and society that do not have much to do with the science of evolution itself.[20]

Indeed, not everyone agreed with this discussion of the rower pool.[21] Elliott Sober and Richard Lewontin[22] suggested that Dawkins' description was an incomplete view of evolution. They pointed out that one must take into account the overall composition of the rower pool to determine which rower type is a good rower. When the winner depends on the composition of the rower pool saying that you have a competition between rowers seems to miss the point. Still, this argument itself does not seem sufficient. After all, it is always true that how well an organism does is relative to the other organisms that are around it.

The problem that we will point out is, in an important sense, more basic.[19] Dawkins made an assumption that he did not discuss, but which has surprisingly far reaching consequences. The assumption is hidden in how rowers are placed into and taken out of the rower pool. He assumed that this would be done at random. What happens if we don't do this? Consider, for example, what would happen if we have a rower pool as a line of rowers. We take rowers out of the front of the line and put them into the boats, then place the rowers that win back into the rower pool at the back of the line. In this case we have a very different process of change of the rower pool than if we just randomly take them out and put them in.

We can see that rowers in a certain place along the line of rowers will tend to become the same type, English or German speakers. However, the type that is found may be different in one part of the rower pool line than another. This is quite similar, if not exactly the same, as the model of children in kindergarten choosing to buy Pokémon cards, or Beanie Babies. We have patches of those that are English speaking and patches of those that are German speaking. In the case of rowers, the boundaries between these patches might move from time to time. Nevertheless, the existence of patches makes many aspects of the evolutionary process quite different from the mixed rower pool.

One way that the evolutionary process is different is that if we start from a mixed rower pool, both English and German speakers continue to exist for a very long time. It might happen that one or the other of them will disappear eventually, but it will take many more generations for this to happen if we have a line as a rower pool than if we mix them up every time. Interestingly, this might also be the reason that there are English and German speaking people. If everyone was mixed around in the world, then it would make sense to have only one language, but if people who speak German live in one part of the world, and people who speak English live in another part of the world, it makes sense to have a patchy language structure, where some areas are English speaking and some are German speaking. Today when people move more than they use to, there is more of a tendency to speak a single language than when people didn't move around as much.

There is another major effect of changing how the rower pool works. This effect has direct implications for the idea of the selfish gene and its lack of cooperation. We can see that if there are patches of altruistic and patches of selfish individuals, the selfish individuals will actually do worse than the altruistic ones.[23] This is because the altruists are generally near altruists, and the selfish individuals are near selfish individuals, the end result being that the altruists will tend to do better than the selfish ones because the selfish ones are unable to take advantage of the altruists as they are not next to them in the rower pool. The altruists benefit from being near other altruists, while the selfish ones suffer from being near selfish ones. Still, there must be boundaries between the two where selfish individuals are near altruists, and ultimately the trick is to understand what happens at the boundaries between types. Boundaries are important for many reasons. Indeed, a kind of boundary—the cell membrane—is for many the essential property of living organisms. While a detailed discussion is beyond the scope of this book, studying the effects of boundaries is an important part of thinking about evolutionary processes once we leave the Neo-Darwinian realm.[24]

The biological analogy of the linear rower pool is the mating of organisms that are nearby to each other, but might be quite far away from organisms that are of the same species (or alternatively selective mating according to traits that leads some organisms to be more likely to mate with others of similar types). It is enough for animals or plants to reproduce near to the place where they were born to change dramatically the conclusions of Dawkins and other Neo-Darwinians. This change might be very surprising to some, but it is extremely important for understanding why competition between genes is not enough to describe evolutionary processes. Thus we see that there are many problems with the gene-centered view of evolution and its one-sided emphasis on competition is just one of them.

The idea of evolution as competition has often been applied to human societies. Notoriously, proponents of Social Darwinism argued that just as in the struggle for survival in the wild, people should compete with each other for success, and we should feel no qualms about losers who do not survive. Variations on this "what works, works" argument were often used to justify the terrible living and working conditions of the poorer classes. That view has become extreme, but many people still retain the sense that competition and cooperation (including the idea of helping others) are irreconcilable opposites. In evolution, the competition between organisms seems to make it impossible for them to cooperate. Companies compete with each other for business. People compete for jobs. Many people will tell you that competition is the basis of the free market system. Politics seems to be about competition for power. The truism "It's a dog eat dog world out there" captures how people often think about the natural and social world. Still, given that cooperation certainly exists, how does it fit into this worldview?

Counter to the traditional perspective, the basic message of this and the following chapter is that competition and cooperation always coexist. People see them as opposing and incompatible forces. I think that this is a result of an outdated and one-sided understanding of evolution. Recent work on evolution has produced a more nuanced perspective in which cooperation and competition must act together. Evolution can still provide a framework for thinking about success and failure that is not based on specific values or opinions about good and bad. This is extremely useful in describing nature and society; the basic insight that "what works, works" still holds. It turns out, however, that what works is a combination of competition and cooperation. We can see how this combination works quite clearly in organized sports.

Chapter 7

Competition and Cooperation

Selection in competitive sports[25]

It is clear that a form of selection occurs in competitive individual sports. In a regional 100-meter dash competition, the top few are selected to win prizes. This often makes them eligible to compete in another race with other regional winners. In a more or less organized fashion, the selection process continues until the world champion is identified from the few best racers from various subregions of the world. Selecting just one individual from the entire world is not really the same as selection in biology, in which there are typically many "survivors" in any one generation that give rise to the next generation.

Heredity also doesn't work exactly the same way in sports as in biology. Aside from the few cases where children of racers are themselves racers, there is no direct biological inheritance of athletic ability. There is a different kind of heredity, however, through transmission of knowledge—knowledge of how to prepare and train, physically and mentally, for competition, as well as how to compete effectively during an event. Biological parent-to-child heredity is therefore replaced by teacher to student heredity. Just as in biology, where selection involves increased reproduction as a measure of success, the process of evolution by selection in sports involves learning by copying or emulating the most successful competitors.

Sports and biology have one important similarity that can contribute many insights about evolution: the fact that there are many different kinds of sports. Each sport, when played at its best, requires a different set of skills and strengths. This means that selecting the best in one sport is not the same as the best in another sport. The same is true in biology, where there are many different environments, and many different resources (for example, different types of food or different places to make homes) in these environments. We call a particular environment and set of resources a *niche*. The existence of many different kinds of niches is the main reason there are many different kinds of biological organisms. When there are many niches that are partly separate but also somewhat connected to each other, competition, and evolution as a whole, can look very different from the simple process of organism selection we discussed earlier.

When people have applied ideas of evolution to societal problems, it's generally been quite difficult for them to realize that the existence of many different ways to succeed changes the meaning of competition in a powerful way. As our society grows more and more complex, there has been an explosion in diversity of possible jobs and professions. Think of how many occupations today—software engineer, laser surgeon, management consultant, web developer, professional women's basketball player—didn't exist 75 years ago. There are more and more different ways to succeed, not just because of new technologies, but also because of new cultural and social trends. The large variety of possible professions requires people who are changing jobs and children growing up to carefully select what they want to do with their lives. Many people try different jobs, or move from job to job throughout their careers. One of the problems that individuals face in a complex society is not simply how to win the race, but also how to figure out what race to be in.

Competition and cooperation in team sports

Selection has a different kind of structure when we look at team sports, and this is where we can see how cooperation comes in. In general, when we think about the conflict between cooperation and competition in team sports, we tend to think about the relationships between the players on a team. We care deeply about their willingness to cooperate and we distinguish cooperative "team players" from selfish non-team players, complaining about the latter even when their individual skill is formidable.

The reason we want players to cooperate is so that they can compete better as a team. Cooperation at the level of the individual enables effec-

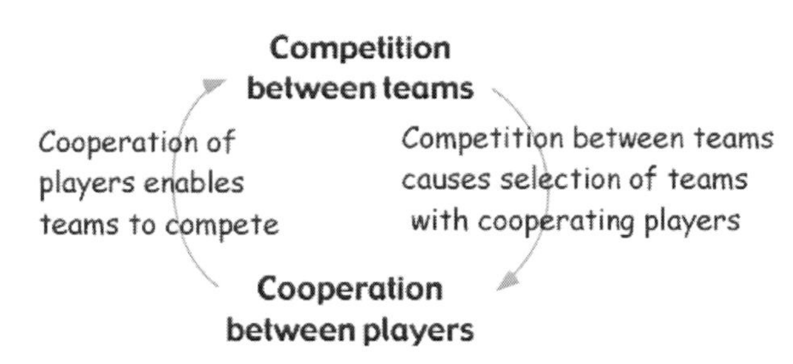

Figure 7.1: Competition and cooperation at different levels of organization.

tive competition at the level of the group, and conversely the competition between teams motivates cooperation between players. There is a constructive relationship between cooperation and competition when they operate at different levels of organization, as conveyed in the arrows of Figure 7.1 that make a cycle.

The interplay between levels is a kind of evolutionary process where competition at the team level improves the cooperation between players. Just as in biological evolution, in organized team sports there is a process of selection of winners through competition of teams. Over time, the teams will change how they behave; the less successful teams will emulate the strategies of teams that are doing well. Teams also change by choosing and trading players and switching coaches. Since the most effective teams are generally those with players that cooperate well with each other, over time teams will be selected based upon their ability to cooperate internally. Competition of teams causes more collaboration within the team, and the collaboration within the team enables a team to compete effectively. The key to this is that competition and collaboration exist at different levels.

The relationship that people notice as a conflict between cooperation and competition happens when both occur at the *same* level of organization, as in the opposing arrows of Figure 7.2. For example, consider the rivalry that existed between Kobe Bryant and Shaquille O'Neal, basketball players and teammates for eight years. There have been periods of time, especially

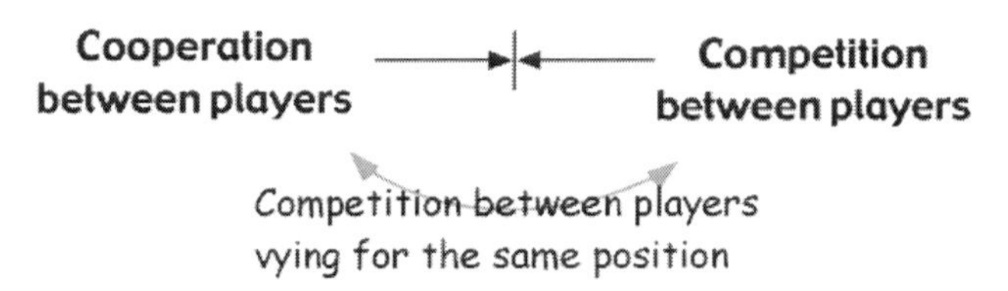

Figure 7.2: Cooperation and competition at the same level of organization.

the beginning of the 2000 season, when these two stars of the Los Angeles Lakers competed with each other for attention. When this happened it conflicted with their cooperation as teammates in a very obvious way; the team did not play well and lost many games. When they cooperated with each other, though, the team was almost unbeatable. On the other hand, sometimes players compete and cooperate without conflict: when those two behaviors are in some sense orthogonal, or independent of each other. For example, players can compete for different kinds of specialized positions (like "forward" and "point guard" in basketball) separately, without interfering with the cooperation of the team as a whole. Because the competition for positions happens at a different time, or between different people, it doesn't inhibit cooperation between players on the team during a game. This works as long as the players are not competing with each other to be the team "star."

Let's step up one level of organization, to the interactions between teams. The competition between teams and collaboration between players reinforce each other, however, teams don't just compete with each other. Professional sports teams are businesses. Basketball teams, for example, cooperate by forming a league (the National Basketball Association (NBA)) that schedules game times, sets rules and enforces them through penalties, selects officials and assigns them to games, and regulates the trading of players. This collaboration between teams is what makes basketball exist as an organized sport. The NBA also competes for media and marketing attention with other sports leagues and other forms of entertainment. All these forms of collaboration maximize profits for the individual teams. Even in non-professional sports, teams must cooperate with each other to decide upon rules and playing times (Figure 7.3).

However, cooperation between teams can sometimes undermine competition between teams. For example, the more games that are played,

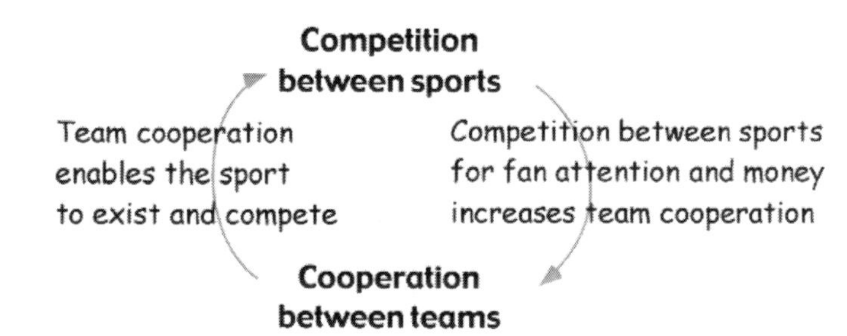

Figure 7.3: Competition and cooperation between teams and between sports.

Figure 7.4: Competition and cooperation between teams.

the more profits are made by the teams, so there is often an incentive in playoff series for the teams to win some games and lose some games. If two teams cooperate in arranging pre-determined victories and defeats to extend a series, this would be counter to the competition in the sport and is considered against basic ethical behavior. Nevertheless, there are people who believe that some sports, like professional wrestling, do this regularly. Whether or not this occurs, in this or any sport, it is clear that there is a conflict between competition and cooperation in this context (Figure 7.4).

Interestingly, in the trading of players, the teams are competing and cooperating at the same time. Teams negotiate to find ways to trade players that each team will agree to, and the potential conflict in doing so is clear—it would seem that one team would gain and one would lose. Still, trading does go on, showing that competition and collaboration at the same time and same level can coexist, even if the relationship is an uneasy one.

The basic point here is this: the interplay between competition and cooperation can only be understood by using a multilevel perspective. Competition and cooperation will tend to support each other when they occur at different levels of organization, but they will generally be in conflict if they occur at the same level. In Figure 7.5 you can see that cooperation at each level enables competition at the higher level of organization.

This interaction between competition and cooperation at different levels has been surprisingly absent from much of the scientific dialogue about evolution. For many years the idea of competition of groups has been taboo because of the focus on selection at the lowest level of organization (i.e., the gene).[26] Why hasn't the interplay of competition and cooperation been central to the understanding of evolution until today? The main problem may be the difficulty associated with visualizing the many levels of organization in a system. It is easier to see one individual organism as the object of natural selection than to see the elaborate schemes of cooperation and competition at all levels of organization. It is even harder to understand when you extend the discussion down to biological molecules. Similarly, when it comes to competition and evolution in social systems, it's difficult to see past the behavior of the individual person to the many

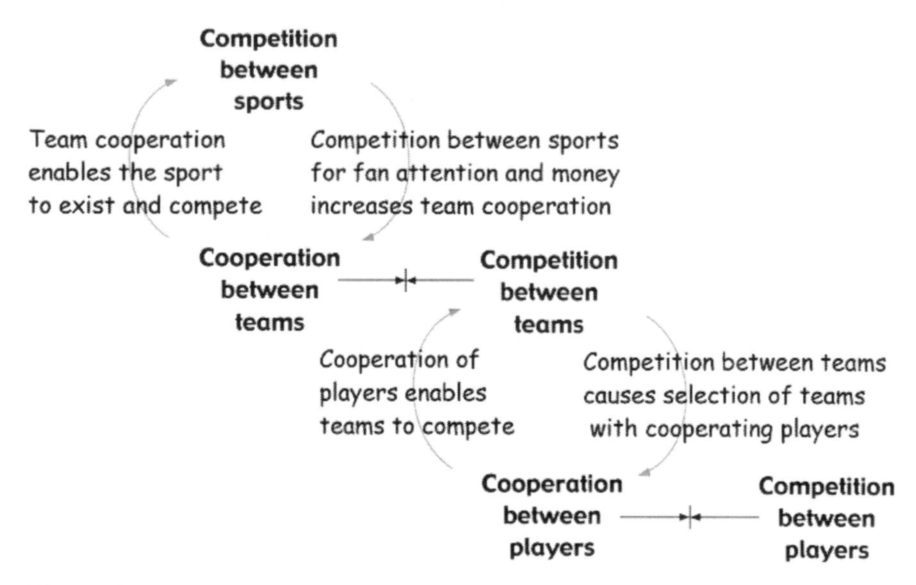

Figure 7.5: Multilevel competition and cooperation in sports.

levels of family, neighborhood, community, social context, country, and so on.

The formation of groups in general—molecules, cells, cells with organelles, multicellular organisms, hives, herds, prides, or other social groups—is one of the key problems in understanding evolution. The traditional perspective on evolution considers the formation of each level of grouping to be a major transition.[27] The use of the word "transition" seems to suggest that this process is *outside* of the more usual process by which evolution takes place. However, the intricate dance at different levels of competition and cooperation in evolution suggests that the formation of groups is a very natural and essential part of the basic process of evolutionary change. Under pressure from their environment, individuals will tend to band together because it makes them more competitive as a collective.

The multilevel perspective on evolution is proving extremely insightful in the study of biological organisms and ecosystems. For our purposes here, however, it's even more important that the multilevel perspective (and the sports analogy) can teach us how to improve the effectiveness of teams in any context. Teams will improve naturally—in any organization—when they are involved in a competition that is structured to select those teams that are better at cooperation. Winners of a competition become successful models of behavior for less successful teams, who emulate their success by learning their strategies and by selecting and trading team members.

For a business, a society, or any other complex system made up of many individuals, this means that improvement will come when the system's structure involves a competition that rewards successful groups. The idea here is not a cutthroat competition of teams (or individuals) but a competition with rules that incorporate some cooperative activity with a mutual goal. For basketball teams, this mutual goal is maximum exposure and profits for the league and the individual teams, and they collaborate through the NBA to ensure that this mutual goal is always in sight. The rules of the game—particularly the ones that prohibit causing serious injury—ensure that the quality of the competition is always maintained. This kind of non-destructive competition happens in nature too, for example, when wild deer and antelope engage in battles over dominance and mating, it is believed that they are careful to avoid severe damage to each other. These battles have real consequences, but they appear to follow rules that prevent mutual injury. We can learn an important lesson about the proper place of rewards for effective competition: the main reward is simply the right to stay together. This, after all, is what survival of a collective is all about.

PART II:

SOLVING PROBLEMS

Chapter 8

Solving Real World Problems

In 1996 the New England Complex Systems Institute was formed primarily to help facilitate collaborations among faculty members engaged in research projects in the interdisciplinary arena of complex systems research. As part of this effort, we organized the International Conference on Complex Systems. At about this time, my textbook, *Dynamics of Complex Systems*, was published. Through these events and the organizational efforts surrounding them, a new and unexpected opportunity arose as we received inquiries on behalf of organizations that were addressing "complex problems." They were interested in understanding the implications of complex systems ideas for their specific problems. Some were also interested in the field's lessons for what to do quite generally, e.g. for organizational effectiveness, organizational survival, or competitive advantage, which often boil down to the same issue in today's complex environment.

In order to address this need, I taught and discussed the concepts of patterns, complexity and evolution with a wide variety of audiences, sometimes in collaboration with other NECSI faculty, other times by myself. The audiences have included members of the intelligence community, the World Bank, military planners, MITRE (systems engineers for the government, especially the military), and corporate executives, often from health care organizations. It became clear that basic concepts of complex systems could be useful to people in their efforts to solve a wide range of problems.

When teaching these concepts, I was also interested in learning about the problems that they were facing, and to see for myself what complex systems concepts could tell us about these problems. In each case I found that direct application of complex systems concepts provided important and immediately relevant insights. Sometimes these insights were new to the professionals in the field and were eagerly welcomed. Other times, they were already known, but the new perspective provided improved clarity. In all cases, the perspectives provided by complex systems research were useful to experts who knew their own profession or business context much better than I.

The chapters in this section of the book illustrate how we can use complex systems ideas to help us solve major societal problems. They can also be considered as case studies demonstrating the wide ranging utility of these ideas, ideas that are useful in solving many other problems that are not directly discussed. It is not my intention to urge everyone to stop what they are doing to help solve all of the problems we face. Instead, there is a need to educate broadly about key ideas that are relevant in practice to understanding how collectives of people can solve problems.

Professionals engaged in one of the areas discussed are encouraged to read through the other topics as they are often cross relevant. While this is counter to professional specialization, we have found that it is helpful to see the ideas applied in more than one area in order to understand how to use them. Moreover, the chapters are constructed so that they build upon each other to develop the set of ideas through all of the applications, rather than each one being a self-contained description of the application of all of the ideas.

It is important to recognize that human civilization is remarkably capable of solving its problems. If there are methods for solving them, they will be discovered and used in many places by many people. Therefore there is no doubt that the recommendations that will be made are already being used in practice somewhere, and generally in more than one place. In this sense, this book is intended to expose or reveal methods that are already being applied, so that they will be used more broadly. Still, not all ideas that are found in practice or combinations of them are correct or constructive. More importantly, people are frequently struggling because they often apply ideas they were taught in school. Many of these ideas are no longer useful due to the changes that have occurred in the world in recent years. I hope that by providing a direct connection between scientific concepts and practical insights, we can clarify which approaches are helpful and which are not. This will enable us to make more progress both in solving these major problems and in recognizing the value of these scientific concepts and methods to a broad range of problems.

A brief review/preview of the essential concepts used in these chapters may be helpful:

When we think about the actions that people are taking as a group, we must think how they interact to form patterns of collective behavior, not just how each individual acts. The patterns that arise result from the structure of interactions between individuals and benefits can arise from both connections and disconnections between individuals. Connections lead to similar or coordinated behavior, which is important when the task involved requires such coordination. Separation and a lack of communication are important when there are independent, or partly independent subtasks to perform. It is easy to underestimate the benefits associated with having the correct kinds of connection and independence because these benefits arise from the collective patterns of behavior that are often hard to recognize.

The most basic issue for organizational success is correctly matching a system's complexity to its environment. When we want to accomplish a task, the complexity of the system performing that task must match the complexity of the task. In order to perform the matching correctly, one must recognize that each person has a limited level of complexity. Therefore, tasks become difficult because the complexity of a person is not large enough to handle the complexity of the task. The trick then is to distribute the complexity of the task among many individuals. This is both similar and different from the old approach of distributing the effort involved in a task. When we need to perform a task involving more effort than one person can manage, for example, lifting a heavy object, then a group of people might all lift at the same time to distribute the effort.

Distributing complexity has some similarity to distributing effort, in that if we want the task to be done successfully, we can't allow any individual to have too much to carry because the failure of one will lead to a cascade of failure throughout the rest of the system. We might then look for someone who can carry a little more weight, but if the way we balance the weight among many people is flawed, there will always be a weak point. Moreover, even the strongest person won't be able to lift his/her share if the overall weight is a hundred times what a normal person can lift and the weight is not carefully balanced among at least a hundred people.

Distributing complexity is different, however, in that no one person can figure out how to coordinate the joint effort of multiple individuals.

The problem is particularly severe because people don't recognize the relevance of complexity in performing tasks. When there is a problem with performing a task, for example, when the existing system fails, the natural way people react is to assign blame to an individual or a particular process and to make someone responsible for fixing the problem. This issue comes up in many circumstances, and many of the central problems in society today can be readily traced to inability to recognize complexity and how it affects us every day.

While some people look to centralized control and individual responsibility to solve complex problems, others look to computer-based automation to solve them. However, it is clear that small children can perform much more complex tasks than computers. For example, many sophisticated people are trying to program computers to recognize the words in spoken language, yet this is a task that children perform easily. Dreams of computers replacing human beings have been around since they were invented, if not before. In the meantime, tasks have become more complex than human beings, not less, and computers, while closer to being able to decipher human speech, are not candidates for this work. This does not mean that computers are unimportant, but unless we recognize the kinds of tasks they are capable of performing, we will be looking in the wrong place for the answers to our problems.

What is the solution to coordinating people to perform complex tasks? Analyzing the flows of information and the way tasks are distributed through the system can help. Ultimately, however, the best solution is to create an environment where evolution can take place. Organizations that learn by evolutionary change create an environment of ongoing innovation. Evolution by competition and cooperation and the creation of composites of patterns of behavior is the way to synthesize effective systems to meet the complex challenges of today's world.

These general concepts find expression and application in the examples described in this section—in complex military challenges, health care, education, systems engineering, efforts to promote development in the third world and efforts against terrorism and ethnic violence. In each case, the complexity of tasks requires us to distribute them among multiple people in order to succeed. For each of the chapters, I have indicated the origin of the discussion, typically a lecture to a particular group, which motivated that chapter. The presentations developed for these lectures have subsequently been used in a wide range of academic, professional and popular contexts and in many venues around the world. I hope you too will benefit from these case studies. Even if the particular field you work in is not addressed, the concepts developed should be useful no matter what you do.

We will begin by considering applications to military conflict. This is a useful first case study because it illustrates clearly the concepts of scale and complexity. In the military the ideas of complex systems are being used in practice through the existence of different organizational structures, (e.g. tank divisions, infantry, Marines, Special Forces) to deal with different kinds of tasks—an extremely important lesson. Much of the rest of the book will be devoted to explaining the value of using this lesson in other social contexts, including health care and education. Moreover, there are dramatic successes of the application of this knowledge as seen in the Gulf War and War in Afghanistan. Still, the military has not formalized its knowledge, and this contributes to the likelihood of such major problems as the current situation in Iraq. Indeed, in Iraq we have created a classic situation of mismatch of scale and complexity on different sides of the conflict. This kind of mismatch is at the core of all of our complex problems.

The second and third topics, health care and education, both have problems with entanglement of scale and complexity, and misapplication of central control. In health care large scales show up in the form of large financial flows, and complexity in the care provided by individual physicians to individual patients. In education large scales appear in the form of standardized testing, while complexity is apparent in the task of preparing individual children for their very different roles in a complex society. In each case the mismatch is leading to problems that require fundamental changes in the structure of the system in order to enable the system to become effective. In paired chapters on each topic we discuss the specific issues of scale and complexity from the perspective of the entire system, and from the perspective of the local functioning of the system: teams of individuals performing a health care task, and teachers and students in individual classrooms.

In the fourth topic, discussing efforts to promote development in the third world, we also have a problem with a mismatch of scale and complexity. Large scale financial aid is not well matched to the multiscale structure of both the poorly functioning societies that exist, and the functioning societies that we would like to create. At the same time, the topic of development also gives us an opportunity to discuss other issues including the role of the environment and external interventions in system creation. Finally, we discuss the problems with planning complex systems that are likely to be hampering the current development strategy of the World Bank.

The fifth topic, the engineering of complex systems, focuses more directly on the failure of central planning. The ineffectiveness of central planning is generally an issue in all of the subject areas. However, engineers have developed planning methodologies to a high art. Since engineers often have more control over what kinds of system are being built, it would seem that they might be successful. Nevertheless, there are remarkably many failures of systems engineering efforts that demonstrate how and why planning fails. In the chapter discussing engineering, we discuss how to replace planning with a different universal strategy: creating an evolutionary context. This approach applies to the creation of effective systems in all contexts, including the military, health care, education and international development.

In the final topic, a discussion of ethnic violence and terrorism, we discuss the challenge of combatting a highly complex terrorist network. However, we argue that this challenge is strongly coupled to a large-scale global movement involving societal changes affecting billions of people. These issues are manifest in many local conflicts manifesting ethnic violence. Recognizing the large scale aspects of these problems is then the key to solving the highly complex problems with both terrorism and ethnic violence.

Together these chapters describe how we can use our ability to understand systems on multiple scales and evolution to address complex problems. In the concluding chapter we will review the concepts, examples and recommendations, and also present a number of examples of successful implementations of these ideas, particularly the use of an evolutionary context.

Prelude:

Military Warfare and Conflict

In 2000, Jeff Cares, then a lieutenant commander in the Navy, invited me to speak to the Chief of Naval Operations Strategic Studies Group, often known as the SSG. The SSG is concerned with generating innovative warfighting concepts (often appropriately considered revolutionary) that are to be implemented over a 20 year period or longer. Each year it is given a task (tasked in military speak) to address a theme posed by the Chief of Naval Operations, the head of the Navy. The SSG is located in Newport, RI, on the site of the Naval War College, part of the extensive educational system of the Navy used for training officers at various ranks. The SSG is not a single group of people with permanent roles; instead, every year a new group of rising officers from the Navy, Marine Corps and Coast Guard come together for a one-year period. They embark on an intensive educational and research program and provide a written response to the CNO and other Department of the Navy leadership. Looking back at previous SSG reports starting from the mid 1990s reveals the clarity with which they anticipated the centrality of asymmetric warfare, threats of weapons of mass destruction and terrorism. Recognizing the importance of networks, they proposed FORCEnet, a networked military organization.

My lecture to the SSG in January of 2000 began a relationship that has involved me in the educational process of all SSG groups since then. I appreciate the interactions during this period with Admiral James Hogg, USN (Ret), Director of the SSG, William Glenney, Deputy Director,

John Dickmann and Jeff Cares. My responsibility included introducing concepts and principles of complex systems and elaborating on them for specific military concerns of the SSG. In the first year, for example, the concerns included perspectives on new roles of the Navy in combat at the interface of land and sea (the littoral region). In addressing this and other questions, the tools of complex systems proved admirably suited for clarifying central issues. What aspects of traditional military thinking, organizational structure and doctrine are important to keep in making innovations in the military? What changes should technology bring in military operations? What approach should be used to implement innovation to transform the organization of the military to the desired new ways? The chapter included here on the military reflects only a small portion of these issues, but a central one: how should organizations be structured to meet complex challenges? It is important to note that the military is typically far beyond other organizations in recognizing the implications of complex systems knowledge from a practical perspective. This has elevated the level of dialogue. Thus in the military chapter, in addition to explaining the role of scale and complexity and organizational structure, we begin to elaborate on the concepts of distributed control and discuss two distinct organizational structures that are both distributed non-hierarchical control systems. These distributed structures have very different properties and capabilities from each other.

Chapter 9

Military Warfare and Conflict[28]

Introduction

Why was the U.S. war in Vietnam so different from the other wars in our past? And why didn't the U.S. win that conflict despite having massively more manpower and a supposedly more developed military than its opponents? There are many arguments for why the U.S. was unsuccessful in Vietnam: the lack of will to win, ambiguous objectives, local economic conditions and global political factors. Regardless of the specific reasons, Vietnam, and other conflicts like the Soviet war in Afghanistan, taught the U.S. military a lesson: It is extremely important to recognize the difference between a traditional, large-scale war and a complex war.

Some conflicts can be won with brute force and conventional frontal attacks. The combatants in these wars can be assessed by quantities like "manpower" or "firepower," and represented on maps by color-coded arrows representing the uniform movement of large masses of troops. (Figure 9.1) When analyzing frontal confrontation of easily distinguished forces in simple terrain, it is pretty easy to see how things are going to turn out. The side with the most troops—and the most weapons—will usually come out on top. The 1991 Gulf War is a recent example. Iraqi forces invaded Kuwait in August 1990. U.S. forces had a clear objective: to drive Saddam Hussein's army from Kuwait. After spending four months building up military "strength" (assembling a large-scale allied force of over

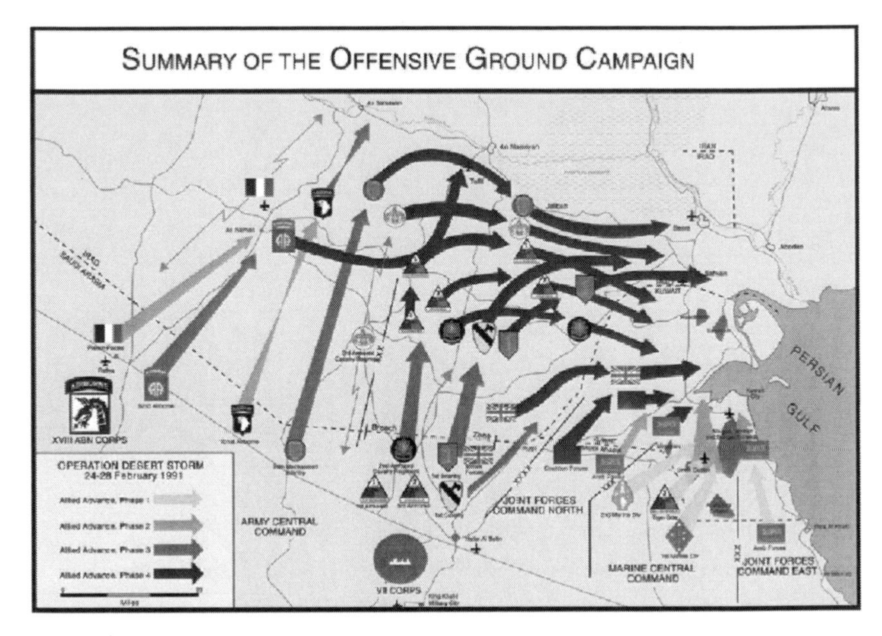

Figure 9.1: A map of the allied ground campaign in the 1991 Gulf War. (from http://www.gulflink.osd.mil/theater.html)

half a million troops) in the region, the allied forces began their attack. Less than two months later (and only a four day ground campaign), the war was over, the Iraqi army soundly defeated.

Vietnam was a very different kind of war, not only in its significantly greater duration. U.S. forces retreated from Vietnam after many years of unsuccessfully grappling with difficult terrain and climate, the difficulty of distinguishing friend, enemy, and bystander, and the inability to locate and target the many nearly independent parts of the enemy. The location of all forces—U.S. and enemy—were never fully known to either side and were spatially mixed. A representative map of what was happening would have depicted the movement of small units or even individuals, in overlapping areas, potentially moving in any direction as they engaged in local conflicts. Lessons learned in Vietnam were central to military effectiveness in the recent U.S. war in Afghanistan, where the mountainous terrain and decentralized organization of the opposition offered challenges that the U.S. was ready for the second time around.

Even more extreme is the current "war on terrorism." The terrorist networks the U.S. wants to disable are dispersed across the world, well-hidden and nearly indistinguishable from civilians. Unlike a traditional army under the command of a single authoritarian leader, this enemy functions in small, independent units only loosely coordinated with one another. In order to overcome this challenge, military and intelligence operations have to be properly organized, so that they will be able to identify and target these elusive units.

These examples begin to illustrate the distinction between conventional conflicts, which are large scale but relatively simple, and complex military encounters. In recent years, the military has recognized that war is a complex encounter between complex systems—systems formed of multiple interacting elements whose collective actions are difficult to infer from the actions of the individual parts.[29] War is particularly complex when the targets are hidden, not only by features of the terrain like mountains or caves, but also by the difficulty of distinguishing among friends, enemies, and bystanders. It is also complex when the enemy is divided into diverse, versatile, and independent targets; the actions that need to be taken are specific, and the difference between right and wrong actions is subtle. Complex warfare is characterized by multiple small-scale hidden enemy forces. Large-scale warfare methods fail in a complex conflict.

The distinction between conventional large-scale conflicts and complex military encounters is now well-known to military planners. Successful military operations, such as the War in Afghanistan, which involved a major role for Special Forces, manifest this understanding through the effective matching of military forces to the nature of the conflict. Still, it is not always clear how to design, plan, and execute military operations to deal effectively with complex conflict. The notions of complexity and scale clarify the intuitive understanding that exists and can offer guidance in selecting appropriate forces for complex encounters, evaluating the capabilities of enemy or friendly forces, and estimating the likelihood of success for specific missions or the overall outcome of military conflict.

Complexity and scale in warfare

Why does complexity matter to warfare? Because war, like so many things, depends on successfully completing planned missions and successfully responding to unforeseen events. A complex mission is one that has a large number of possible unsuccessful actions. The higher the likelihood that a wrong outcome will occur, the higher the complexity of the mission. As we've seen before, a high complexity task requires a system that is sufficiently complex to perform it. So in warfare, as other complex problems, the number of possible actions that the system can perform (and select among) must be at least equal to the number of actions required by the complexity of the task.

Complexity increases in military conflict where the application of effective force must be more carefully selected or more accurately targeted, or where the implications of making wrong choices are more severe. Hidden enemies—particularly when commingled with bystanders or friendly forces—present high-complexity challenges. So does carrying out military operations in an urban setting, especially when objectives require minimal damage to buildings and infrastructure. Forces performing peacekeeping functions are also faced with high-complexity tasks. In volatile regions filled with many conflicts, a single wrong action or missed opportunity can blow up into disaster.

Scale is also an important factor to consider in designing effective military operations. Scale refers to the number of parts of a system that act together in a strictly coordinated way. Imagine a battalion of the army, composed of as many as 1000 people, delivering multiple shots coherently at a single large target (i.e., a uniformly approaching enemy force). Here, the battalion is applying the greatest aggregate force it can, performing its largest scale of action. Or imagine a supply line transporting a lot of equipment from one place to another along a road. The individuals in this group all have to move at the same time in the same direction, thus their coherent action can be seen from miles away.

However, it is not difficult to imagine a scenario in which a force of the same size has to be able to direct the same quantity of firepower at a set of separately specified targets. Some missions call for completely coordinated actions, whereas others call for multiple partly independent actions. Every mission requires a certain scale of effort—measured, for example, by the number of people who have to work together to perform it, or alternatively by the distance from which it can be distinguished as a separate action.

Most systems, including military ones, must be prepared to deal with many tasks on many different scales. So a successful organization is one that exhibits sufficient complexity at every scale of action necessary to complete its required tasks. This is where the complexity profile, which we discussed earlier, comes in—complexity varies with scale differently in every organizational structure. The complexity profile characterizes the dependence of complexity on scale of a particular system or organization. At a given scale, how many possibilities for action does the organization have?

Let's take a look at a hierarchical military structure first—a battalion, for example, successively subdivided into companies, squads, fire teams, and eventually individuals. When forces are organized hierarchically like this, the number of possible actions at a small scale increases as the number of small units (e.g. fire teams) increases. The number of possible actions at a large scale increases as the number of larger units (e.g. battalions) increases. So the complexity profile roughly corresponds to the number of units at each level of command (individual, fire team, squad, company, or battalion).

However, the complexity at a certain scale doesn't just depend on the number of units operating at that scale—it also depends on how independent the individuals are within fire teams, how independent fire teams are within squads, how independent squads are within companies and how independent companies are within battalions. When the units at a particular level of organization are more independent, the complexity is higher at that scale, but as a result, complexity at the larger scale is lower because it's more difficult to carry out coordinated actions. The dependence of the complexity on the scale, i.e. the complexity at the individual, fire team, squad, company, and battalion levels of organization, is the complexity profile of the entire military force.

A force trained and organized to apply large-scale force effectively is not well suited to acting on a smaller scale and vice versa, a force designed for complex small-scale conflicts is not well suited to large-scale conflicts. This trade-off between differently designed forces is recognized by military planners, but often in an anecdotal or ad-hoc way. The complexity profile gives us a way to formalize these discussions in order to evaluate the effectiveness of force design in the face of a specific complex military mission or conflict. It turns out that for the same set of components organized in different ways, the area under the curve of the complexity profile will always be the same.[12] This means that for a given a set of resources—troops, weapons, technologies—we can compare different organizational structures and their capabilities and limitations.

In addition to the nature of the objectives and the organization of forces, the environment itself can add complexity to a conflict at different scales, and using the complexity profile here is helpful as well. The simplest battle space is the ocean—vast, open, and flat. Warfare that takes place on the open ocean mainly takes the form of simple large-scale direct confrontations. This is why the largest military transport structures that exist have been designed for ocean use: aircraft carriers and other large ships.

These massive craft are designed for simple large-scale actions and are not well-suited for responding to threats by smaller enemy vessels, especially hidden ones like mines or submarines. This is why large vessels are usually accompanied by several smaller vessels, which are more capable of detecting and eliminating smaller-scale threats.

In contrast, the interface between land and water (the shoreline, or "littoral" region) is complex at many scales. A lot of information is required to describe the natural features at this interface; there may be coastline, cliffs, marshes, swamps, mud, brush, sand, pebbles, dunes, reefs, or rocks, and their specific shapes and arrangements are also important factors to consider. Operating effectively in this interface between land and water requires being able to move and function well in both. The civilian populations living in this environment also adds to the overall complexity. Cities, ports, transportation networks, land vehicles and boats present many possibilities for action and concealment on the part of friendly and enemy forces.

In a harbor or on a rocky coastline, there are many obstacles that prevent mobility of large objects, such as ships designed for the open ocean. In contrast, small objects such as little boats, pedestrians, swimmers or divers, can maneuver and remain hidden. The attack on the USS Cole on October 12, 2000 was possible because a maneuverable dinghy was able to approach a large ship. The ability of the ship to defend itself was inhibited by the possible confusion of enemy and friend, and by the likelihood that firing might inflict damage to non-enemy structures. There are many ways to attack a large ship in the littoral region—and if collateral damage is to be avoided, there are few defensive and offensive actions that the ship can perform when confronted with many or even a few small enemies. This is a weakness of large-scale warfare in complex regions like the littoral, but a strength for small, independent, highly capable individuals or small groups of individuals that can use the complexity of the terrain to their advantage.

The complex physical environment does not determine by itself the complexity of littoral warfare. The environment is the context in which military forces engage in their tasks. This means that the environment often becomes an important part of the challenge, but is not the challenge itself. The complexity of the terrain can be used to enable an attack, and it limits the effectiveness of forces that are not appropriately structured. As the case of the USS Cole demonstrated, a small, even low-technology force can effectively attack a much larger force in a complex environment.

The difference in complexity between land-water interface and the open ocean is already manifest in the radical differences in organizational structure, training and equipment used by the Marines, as opposed to the Navy. The Marines, who were originally created to operate in this terrain, embody many of the specific implications of littoral complexity, especially the need for small independently acting groups and more distributed control. Highly reliant on individual training, the Marines are known for the diverse, resourceful and specialized nature of their individual and group forces. They also make extensive use of technology that enables functionality in a complex environment. In general, an effective force in the littoral will be one that allows individuals or individual teams to function effectively in the local context with limited coordination between units.

More generally, the complexity profile can be used for identifying what kinds of organizational structures are suited to specific terrains. Matching the complexity profile of the force to that of the environment enables it to be effective. This is manifest in the structure of existing military forces. Large ships are well suited to the ocean, the simplest terrain. Tank divisions are well suited for deserts and plains. Heavy and light infantry are suited for progressively more complex terrains, such as towns, fields and forests on increasingly hilly land. The Marines, with their small fighting units and high levels of individual training for independent action are suited for the interface of land and sea. This might seem rather obvious—of course tanks can't move effectively in the mountains (or in a dense jungle)—but the reasoning behind this runs much deeper than the ease of transportation in different terrains. The individuals in tank divisions are trained and organized in command protocols to act on a larger scale than the Marines are, and their characteristic scale of action is not well suited to terrains where finer-scale actions are necessary. Forces cannot be designed for success in both large scale and complex terrains. In complex terrains, Marines will defeat infantry, infantry will defeat tanks, and tanks will defeat ships. Even a single Marine can defeat many ships in the complex terrain near a shoreline.

The experience with highly complex warfare in Vietnam led to the creation of even finer scale "Special Forces," including the SEALs, Delta Force, Rangers, and Green Berets.[30] These forces are organized as small, highly trained teams. Some of the Special Forces are trained to act as tightly coordinated units, while others act more as a collection of individuals in specialized roles. Their training is not only for specific military battles but also for gathering information and developing relationships with other military forces or civilians. Special forces are not designed for the largest scale force. They are designed for highly complex conflicts. Their effectiveness in the recent War in Afghanistan demonstrated how the climate difference between the jungles of Vietnam and the mountains of Afghanistan was not as important as the similarity in need for small independent teams and highly individualized training.

Command and control structures

Network vs. hierarchical control

Using the complexity profile, we can begin to understand the limitations of hierarchical command. In the first section we discussed some of the problems associated with an idealized hierarchy. To the extent that any single human being is responsible for coordinating parts of an organization, the coordinated behaviors of the organization will be limited to the complexity of a single individual. Using a command hierarchy is effective at amplifying the scale of behavior, but not its overall complexity. Therefore, hierarchies are ineffective at performing high complexity tasks.

By contrast, a network structure (like the human brain) can have a complexity much greater than the complexity of any of its individual elements (neurons). While an arbitrary network is not guaranteed to have a complexity higher than that of an individual component, it is possible for such a network to exist. For high complexity tasks, we therefore consider hierarchical systems inadequate and look to networked systems for effective performance. This explains the recent trend toward distributed control in corporate management, as businesses discover the limitations of hierarchical organizations in the face of the modern complex socio-economic system. Network warfare concepts are also gaining influence in current military thinking.

It is important that people are beginning to break free from the traditional notion that the only alternative to hierarchical control is anarchy. The concept of a network as a model of social and technological organization is now in widespread use, and is usually used to suggest widespread availability of information and coordination. However, distributed control is not a panacea for the problems associated with hierarchical control and it won't lead to more effective systems in and of itself. In fact, "distributed control" doesn't actually correspond to any specific control structure. The capabilities of a distributed network must be more carefully understood in relation to the function it's supposed to fulfill. Only a control structure that is effective for the specific tasks at hand will ultimately prove successful.

Two examples from human physiology

What kinds of networks exist that would be useful for military organization? Nature provides a variety of examples that we can analyze for their relevance. It might be helpful to start with two examples from human physiology: the immune system and the neuromuscular system.[31] The immune system consists of a variety of types of agents (cells), many of which are capable of movement, have sensory receptors, communicate with each other, and are capable of attacking harmful agents (antigens) as part of the immune response. These agents act independently, but achieve some degree of coordination of activities and functional specialization through communication. The immune system's complexity profile tells us that the system acts with high complexity at a very fine scale, with many independent agents acting differently at a given moment in time. The individual actions of those agents rarely aggregate to large-scale behaviors, so the immune system does not have high complexity at scales significantly larger than the scale of cellular action.

The neuromuscular system provides a very different example. It's composed of two segregated components: the distributed network of neurons known as the nervous system (in which are included the senses), and the muscles that consist of highly synchronously (coherently) behaving muscle cells. The central decision-making component of the nervous system, the brain—also a distributed network—processes the information from disparate sources into decisions about action. At any given moment in time, the neuromuscular system is performing only one or a few individual large-scale actions—muscular movements, such as lifting an arm or taking a step. These motions are visible from a great distance, unlike, say, the motions of white blood cells, which can only be observed with a microscope. Still, the human body exhibits complex neuromuscular behaviors over time because each action is selected from a variety of possible large-scale actions—lifting an arm more slowly or at a slightly different angle, for example, or taking a step in another direction at a slightly different speed. The nervous system's distributed network allows for the selection of the variety of possible large-scale actions, using sensory information. As a whole, then, the neuromuscular system selects from a variety of actions at the large scale (at the level of macro-muscular movements) when considered over time.

The complexity profile of the neuromuscular system is very different from that of the immune system, which cannot produce large-scale behaviors. The immune system has very complex behaviors at the cellular level—not only over time but at any given moment of time. The differences between the complexity profiles of the two systems make sense, as the immune system's function is to protect its host (the human body it resides in) from disease and infection internally. Only when foreign agents enter the internal environment of the body does the immune system respond to them. Its complexity on the fine scale allows it to effectively fight the tiny bacteria, viruses, microbes, toxins, or parasites that infiltrate its host. In contrast, the neuromuscular system responds to external objects or conditions on the scale of the body itself. These objects are separated from the body by a margin of space that is typically larger than the body. The neuromuscular system's ability to generate complex behaviors on the scale of the entire body is crucial to the body's survival in its macroscopic physical environment. The neuromuscular system is as useless in defending the body from viruses as the immune system is in preventing a car door from slamming on your finger.

This comparison between the immune and neuromuscular systems begins to demonstrate how certain organizational structures are effective for certain environments and tasks. It also illustrates the importance of functional segregation—both the immune and neuromuscular systems are specialized subsystems within the same organism, one suited to protect the body's internal components and the other suited to respond to the external environment. This comparison reiterates the key point that organizational structure always reflects a trade-off between scale and complexity and the organizational structure must be related to the system's function. A system designed for complex large-scale behavior has a very different structure from a system designed for high(er) complexity behavior at a fine scale.

These two physiological systems are useful as models for military organization precisely because of their differences in function. The immune system is effective at carrying out many localized and simultaneous tasks, whereas the neuromuscular system is effective at determining a single, but highly selective act at any one time. An effective military can utilize both types of organization, but must recognize that each requires significantly different organization, training and technology.

Networked action in warfare

A military organization modeled on the human immune system will be a system of largely independent agents. Each agent will be capable, on its own, of sensory activity (observation and reconnaissance), decision-making (analyzing information to select an appropriate action) and action. Such a versatile "agent" could be a single highly capable warrior or a small, tightly connected team of warriors with diverse training or equipment. Because they are capable of acting without being controlled or directed from above, this kind of agent is an 'action agent' and by allowing multiple agents to interact we can have networked action agents, which also may be described as distributed action agents.

Distributed action agents in the same vicinity will communicate with each other to coordinate local actions, but will be mostly unconnected to forces further away from them. They interact with each other to coordinate their individual actions for effective attack, defense, search or other tasks. This coordination allows them to achieve the level of local capability needed for a task. When one or a few individuals are necessary for a particular task, others should not congregate there; when more are necessary they should.

The local coordination will be different for every mission and environment. Complex conflicts tend to have distinct local conditions and ground warriors securing a jungle region may have different communication methods and needs than a group of small watercraft intent on disabling a docked ship in a guarded harbor. In some cases, simple shouting and signaling with hands will do; in others, more technologically advanced means will be necessary.

Consider the following example. When crossing through a barrier of rough or changing terrain, such as might occur when landing on a rocky coastline, or passing through a treacherous stretch of mountainous terrain, a simple but efficient means of communicating the location of passages ("it's easier over here") will allow more effective movement, especially when visibility is limited or secrecy is crucial. There would be little point in setting up a centrally coordinated movement here because the terrain is so varied. Knowing the location of an access route is only helpful if it is nearby, so local communication will be far more effective. How successful would the 1944 American landing on Omaha Beach in Normandy have been if commanders had attempted to coordinate movements by shouting directions through a loudspeaker from the battleships? These directions would have been near useless to individual soldiers dealing with the uncertain footing, mined obstacles, and German fire within the 10 or 20 feet around them.

Instead, distributed action agents using local coordination of sensors, movements, and fires to achieve a larger scale effect than is possible with a single warfighter, is what is helpful in the above scenario. No matter what the actual communication protocols, this local coordination will lead to the formation of simple collaborative patterns of movement. The Normandy landing is often described as Allied forces "swarming" ashore, and the metaphor is actually very insightful.[32] In a flock of birds or a swarm of insects, for example, the individuals adjust their motion only in response to the positions and speeds of the few other individuals around them, but this simple coordination generates motions on the level of the entire swarm that adapt quickly to local situations like obstacles or other causes for change of direction. The kind of coordination that forms simple collective patterns, such as flocking or swarming, is very different from the more intricate tactical planning of carefully timed collective actions.

Local coordination thus produces emergent collective behaviors that could never be directly specified through hierarchical control. The specific pattern that arises is determined by the response of the agents to the local challenges they face in the environment as well as through interactions with each other. Efforts to control the local actions globally would inhibit local adaptation to challenges—just as a commander using a loudspeaker to direct the landing on the beach at Normandy would have been a disaster.

The way such emergent pattern formation occurs from local rules of interaction is generally considered fantastic and mysterious; however, one of the goals of complex systems research is to demystify these patterns in order to understand their mechanisms and effectiveness. Interaction rules like "local-activation long-range inhibition," which we've discussed earlier in the book, are key to understanding "mysterious" collective animal behaviors like swarming and flocking. Similarly, if warriors are bound to simple, limited rules of local interaction, their individual actions will form interesting and useful collective behaviors—and that's not mysterious at all!

The simple coordination that is possible through local interactions is a powerful mechanism, but it is not suited to every kind of situation. It is most effective in a fine-scale, high complexity terrain, where independence is important, but some coordination is also necessary to deal with local variations in environment or task objectives. It is essential when the task at hand varies widely from place to place in a way that would overwhelm any attempt at central control.

These simple pattern-forming processes have their limitations since they do not produce elaborate forms of coordination between individuals or groups, and more elaborate coordination is sometimes necessary for effective action. This is when distributed action agents must have a higher level of practiced coordination and exercised teamwork—the kind of team effectiveness that is a key aspect of conventional military training. The trade-off is that the additional coordination limits the flexibility of individual action. This is why the concept of "distributed control" is not precise enough to offer a universal method for improving military organizations. There is a spectrum between weak coordination and large-scale coherence of forces. The important step to take is not to implement distributed control, but to determine how distributed a control structure needs to be in order to successfully face the objectives of the conflict at hand.

Distributed control coherent action

Let's return to the other physiological model: the neuromuscular system. The part of the system that makes decisions about how the body should act (the nervous system) is a distributed control network of neurons that interact with other neurons. The part of the system that actually carries out these decisions (the muscular system) is designed for large-scale impact, producing coherent macro-muscular movements. Because of the networked decision system, the choice of when and which large scale impact to perform can be made highly selectively, based on disparate information sources. The complexity of the human body's movement arises because each act at a particular time can be precise and carefully selected, and different acts can be selected at subsequent times.

A military organization modeled on the neuromuscular system would possess large-scale conventional (or modernized) force capabilities. However, instead of being controlled conventionally (hierarchically), they would be coupled to a highly distributed decision-making process, enabling many factors about the current situation to be considered in the selected action. Having the capability to act at a large scale doesn't mean it will always be fully used—just as having muscles that can kick or punch doesn't exclude using a delicate nudge at times when it would be effective. The force to be used is selected carefully from many options to achieve the desired objectives. Whereas the prime objective for a system of networked action agents (like white blood cells, or teams of Special Forces) might be to deliver fires to many different targets at the same time, a system like this will be best suited to delivering the right force to one particular right target at the right time through a remarkable understanding of the specifics of the entire situation as it changes in time.

This kind of system demonstrates one way that centralized control and hierarchical control are not the same thing. The brain serves as a central decision-making center, but it's also a distributed network of neurons. This is a very different kind of centralized control than one in which decisions are made by a commander receiving information from a few individuals (e.g. those just below him in a hierarchical command structure). Still, in the nervous system there is only a relatively small set of neurons that direct the action of the many cells in any particular muscle. This is consistent with a relatively recent concept in military doctrine (developed by the Marines): distributed control with central command. Centralized command can be consistent with distributed control—and in fact, they work quite well together, as the neuromuscular system demonstrates.

Conclusion

The respected scholar of military command, Martin Van Creveld, once said that war is "the most confused and confusing of all human activities."[33] War is confusing, but it's not so confusing that it can't be understood at all. As with so many other things that seem hopelessly mysterious, complex systems concepts can help to make some sense of the confusing business of war. With these conceptual tools—complexity and scale, the complexity profile, and an understanding of different control structures—it's possible to understand and analyze quite rigorously the behavior of complex military encounters that might have seemed mysterious at one time.

Conventional wars were large-scale challenges where the biggest forces won.[34] In a complex war, the organization of forces is as important as the size of forces. Instead of applying a large coherent force, by bombing and sending in tens to hundreds of thousands of troops, as in the Gulf War, a complex war may require the same number of troops—but organized into multiple weakly-coordinated forces, acting at the same time on various independent missions. Decreasing the scale of coherence of forces may turn out to be the key, which might seem counterintuitive. On the other hand, when large forces are needed, a distributed decision process, involving many networked people with different sources and types of information, may enable us to effectively select the right action to be taken: the right force to be applied at the right place at the right time. In this case the scale of action may be large but the act itself is highly selected. In general, each complex warfare situation must be met by military forces well suited to the conflict.

The discussion of the relevance of organizational structure to effectiveness is not unique to military conflict. The military, however, seems to learn quickly from experience, at least in part because the lessons learned are often so immediate and clear. Another reason for military learning, however, is the existence of a culture that is determined to learn from past experiences with a forward-looking strategic perspective about future conflict. This has led to the creation of military organizations that are clearly built around the needs of different kinds of terrain as well as the structure and strategy of enemy forces. It is important to recognize the generality of these lessons. Complex systems—and particularly the complexity profile—provide a way of understanding the lessons learned from past failures in a way that can be applied much more broadly.

It's very telling that over the past 20 years, the notion of war has been used to describe the War on Poverty, the War on Drugs and other national challenges. These were called wars because many believed these challenges require the large-scale force of conventional wars. However, they do not. They are complex challenges requiring many different actions in many different places. Allocating large budgets for the War on Poverty did not eliminate the problem. The War on Drugs has taken a few turns, but even the most recent social campaign "Just Say No" was a large-scale approach. We have not won these wars yet at least partly because we are using the wrong strategy—and that faulty strategy is reflected in the limited understanding of the metaphor most often used to describe them. War is not always a large-scale affair and neither are our other most complex challenges.

Prelude:

Health Care

In 1998, Helen Harte, a health care quality consultant based in Seattle, offered to organize a session on medical management for the International Conference on Complex Systems. She invited outstanding speakers who together formed a remarkable session. From them we learned that complex systems as a science had become of interest to medical management, because of its potential to help them in their struggles with the complexity of health care. The VHA, a hospital association, had developed programs promoting this interest. Helen continued to be involved in NECSI programs. She organized a public one-day program on complex systems concepts in Seattle that was sponsored by Microsoft, Boeing and Group Health Cooperative. She also suggested the development of a course that would serve executives and senior managers from all types of corporations; and we subsequently organized the "Managing Complex Organizations in a Complex World" course.

The Managing Complex Organizations executive education course, the first of which was offered in the spring of 2001, has been offered twice a year since then. The lecturers for all of the programs included Peter Senge, author of *The Fifth Discipline*, and founder of the Society for Organizational Learning, John Sterman, author of *Business Dynamics*, and head of the Systems Dynamics group at MIT, and myself. It has been an honor and a pleasure to work with Peter and John on this program. As two of the world's most highly respected management experts, their understanding is deep, and their professional collaboration is greatly appreciated. Further, I respect their personal values and caring about people generally and every individual they interact with specifically. Other lecturers in this program have presented highly appreciated lectures in individual courses: author Tom Petzinger, Helen Harte, emergency room director Mark Smith, researcher Ary Goldberger and Jeff Cares, who has become a military consultant and is applying complex systems concepts in this context. Through their work, each of these contributors has had an important impact on the understanding and practice of managing complex organizations. The participation in this course has been quite broad across all industries, but there has been a strong presence of health care executives, reflecting the ongoing turmoil in health care.

The Managing Complex Organizations program was composed of four parts. These parts reflect perspectives that leadership can use to improve their organizations, where leadership is broadly understood to be not only for the CEO or senior executives, but surely includes them. The four perspectives are briefly described as follows:

Part I: Patterns—A metaview—stepping outside day-to-day responsibilities to develop a perspective on the behavior of the organization as a whole. Learning how interactions among members of the organization and information flows determine this pattern of behavior enables us to identify unsuccessful patterns of behavior and how to intervene.

Part II: Possibilities—A metametaview—looking beyond the current pattern of activity to consider the set of possible actions that the organization can make. This view relates the structure of an organization to the capabilities it has, or, given a set of tasks, identifies which organizational forms are capable of effectively performing these different tasks.

Part III: Organizational Ecology—learning how to promote organizational change that will directly impact effectiveness. Specifically, how we can foster an organizational environment that supports rapid adaptation.

Part IV: Leadership—developing a perspective of the self in the context of complex organizations. Learning how the many roles of leadership, including creating objectives and an environment of mutual and collective engagement, are related to organizational performance.

When it became clear that health care executives in particular were very interested in the opportunity to learn from the ideas of complex systems, I considered some of the key aspects of problems in the health care system and developed an understanding of the problems which will be described in the following chapters. It is quite clear that the health care system is facing serious problems: high rates of medical error and low quality of care. The basic effectiveness of individual physicians and the physician patient relationship continues to be remarkably strong. Still, this doesn't address the problems of health care because physicians don't act alone. They act in combination with other physicians and practitioners involved in providing care. There is a need to change our perspective on the entire health care system in order to solve its underlying problems.

At some point, one of the participants in our course engaged me in limited contact with several people who have been concerned with establishing new standards for educating physicians (specifically for educating residents) because of their recognition of the importance of complex systems in the health care context. The new standards they developed have recently come out. The standards include the need for training in "systems-based practice." Systems-based practice is the perspective that care is no longer provided just through the attention of an individual physician, but by the interplay of many people organized in a coordinated system. Training of physicians is traditionally focused on how they as individuals should provide care to a patient, which does not provide the knowledge or skills for working in a system of care involving coordination with others and utilization of the capabilities that others provide. The idea of systems-based practice is to provide this type of information to physicians. This is an important development, however, we need to go even farther in our training. Beyond the training of physicians, we need to think about the training of systems. We are working on both training physicians for systems, and training of systems themselves, as part of a new program at NECSI. In the meantime, it is important to communicate and clarify these issues so that we can take the steps necessary to overcome the crisis that currently exists. If this is only the first step, it is clearly an essential one.

During the past year, we have developed a program, the NECSI Health Care Initiative, in collaboration with the Centers for Medicare and Medicaid Services, the Centers for Disease Control and Prevention, hospital systems and other health care organizations, to work towards developing projects that will implement complex systems ideas in health care and public health systems.

CHAPTER 10

HEALTH CARE I:

THE HEALTH CARE SYSTEM[35]

Introduction

People have been talking about the health care "crisis" in the U.S. for almost two decades—since the rapid growth of managed care began in the face of rising costs. More money is spent on health care in the U.S. on a per-person basis than in any other country in the world. However, if standards can be used as a measure, the quality of health care in this country is far from the highest worldwide.[36] Many complain that the current system provides a very poor return on investment compared to the care in other countries. Other symptoms also point to a system in trouble: a notorious medical error rate and low quality of care. Why is this happening despite the expansion of medical knowledge, the use of increasingly sophisticated technology, and the high level of training for physicians in this country?

The answer, we'll show, lies in the basic financial structure of the health care system. Managed care's efforts to lower costs through industrial era methods of efficiency are incompatible with providing complex individualized treatment. This streamlining approach has been weakening the system's ability to provide effective medical care because it's no longer suited for the high-complexity tasks it performs. The key to understanding this is recognizing the distinction between large-scale and complex.

An important question to ask is why the system isn't providing high quality care automatically. Why doesn't the local system (e.g., a particular hospital) work to fix itself? Shouldn't the system improve the quality of the care on its own? To answer these questions we need to have a broader understanding of the health care system—what the key external forces are, as well as the internal interactions. Over time, the external forces acting on a system affect the changes that are taking place within it and determine a lot about its development. Are these pressures moving it in the right direction (not likely, because then the problems wouldn't be there)? Are they not moving it in any particular direction (possible, but not what we should expect in the current situation)? Are they moving it in the opposite direction (most likely!)?

To understand this, we need to look at the overall structure of the health care system. Of course there are many aspects of the health care system, but there are some key aspects that we can discuss simply, which will help us understand what is going on and why the system is going in the wrong direction from the point of view of individual care.

The financial structure of the health care system

About a hundred years ago, medicine was largely practiced through individual relationships between practitioners and patients, practitioners were not yet as specialized or organized together as they are today. The basic interaction would have looked something like that shown in Figure 10.1.

Through the course of the twentieth century, the development of health insurance and the trend towards managed care have changed this picture

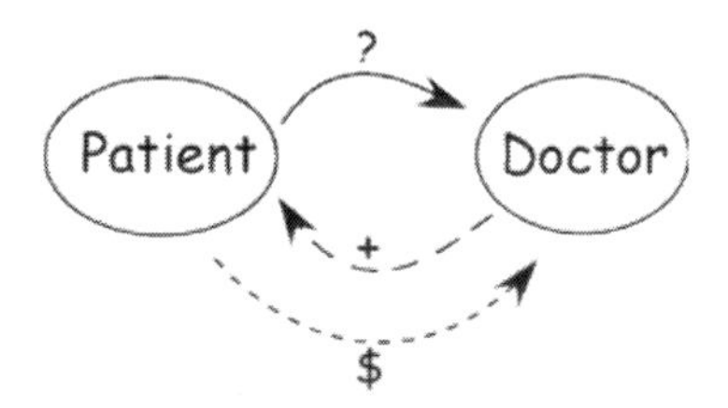

Figure 10.1: The traditional doctor-patient relationship. Information about symptoms, etc. (arrow marked with ?) flows from patient to doctor, and information about diagnosis and treatment (arrow marked with +) flows in the other direction. Meanwhile, money (arrow marked with $) is transferred from the patient to the doctor as payment for treatment received.

significantly. Today, most individuals do not directly pay their physician or other practitioner in full for their services. Payments from patients to doctors, "co-pays," do not cover the cost of medical services. Instead, employers (or, less often, individuals) make regular payments to their insurance companies, other health plans, or Medicare—payments that are not directly related to the actual services provided during that time period. Practically speaking the payment is often an electronic bank transfer once a month. Part of the money may be deducted from employee salaries, while the other part comes directly from the company. Either way, the payment amounts are decided upon in advance and are the same from month to month, until rate changes take place, typically on a yearly basis. With respect to the nature of the actual medical care provided, this sum is essentially featureless: large scale and simple, having no information encoded into it about the complex medical services it will eventually fund.

The insurance company or managed care organization divides this large-scale flow of money into smaller financial flows to the different health care providers in its system. Sometimes they go directly for specific services, payments for treatments to specific physicians. Other times they are paid as intermediate sized payments to health care organizations, which are then allocated as compensation for individual practitioners, or as funding for procedures, supplies, and other medical costs.

The diagram in Figure 10.2 represents the flow of information, services,

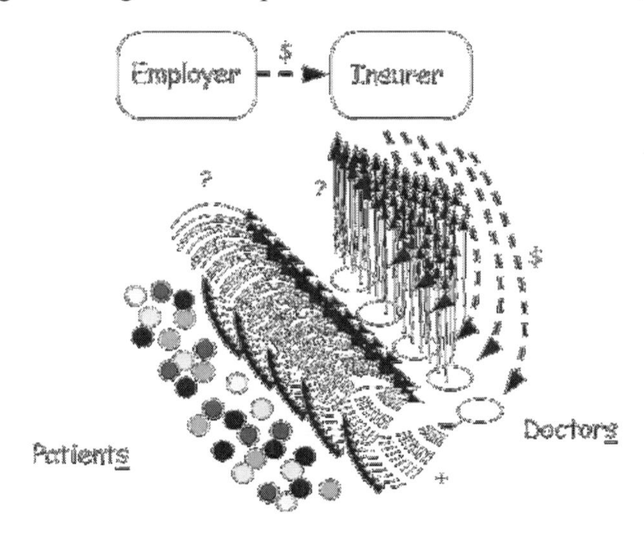

Figure 10.2: The structure of the health care system today. Insurers receive lump sum payments from employers, and use these funds to pay doctors for specific services.

treatments, and money in the existing health care system. Information and medical treatment are exchanged in the transactions between physicians and patients, whereas the flow of money is largely from employers to health care insurers and thence to health care provider systems and individual practitioners. The current health care crisis has its origin in the structure of these flows. Let's take this concept of "flow" seriously, and consider an analogy to a complex systems phenomenon which helps explain why this type of system is not likely to be effective: fluid turbulence.

Turbulence

Turbulence occurs when a simple coherent flow is broken up into many smaller flows. It can be observed in the swirls and eddies in a fast-flowing river, or in the way the coherent column of smoke risings from a camp fire breaks up into swirling patterns as it ascends. Although we can identify situations where turbulence will occur, it's very difficult to predict the resulting motions, which are irregular and change rapidly over time.

In the health care system, we have an analogous situation. The large-scale financial flows that drive the system eventually have to be allocated as small payments to individual doctors treating individual patients for individual problems. The transition from the large- to the fine-scale is turbulent for financial flows just as it is for fluid motion. The idea that turbulence is the analogy to what is going on in the health care system will not come as a surprise to the people who work in it, as they have experienced the turmoil over the past 20–30 years. The unpredictable rapid changes have not been in the relationships between doctors and patients, or in the relationships between employers and insurers (though sometimes they feel involved, at least as interested spectators)—the main changes have been between the insurers and the physicians. The growth of managed care, physician cooperatives, reporting and billing systems, and hospital mergers, are all part of the interface between the insurers and physicians. These changes, particularly the way physicians and hospitals are joining together to create larger groups that provide many medical services, are a response to the unstable flows of money. People are joining forces in an attempt to stabilize and control the flows.

What does this turbulence look like in human terms? The problem of large flows connected to highly complex flows is abstract, but the reality is quite easy to recognize. Eventually the issue is related to the problem of controlling the flow, specifically: Who is making the decisions that control the flow of money in this system? Since the early 1970s and increasingly since then, an effort has been made to control the flow at the large scale end. Companies and insurers, frequently with the intervention of state and federal government organizations, negotiate the rate of flow of the money from employers to insurers. They decide on changes in the rate from one year to the next. How do changes in this rate affect the system? We have to think about how this translates into the flow in the system.

Consider the effects of a simple action, such as changing the flow at the source, by increasing (or decreasing, though practically speaking the former is more likely) the amount by a certain percentage (e.g., 8%). This kind of increase in spending is typically done on an annual basis. The amount of increase reflects a decision about how much should be spent on health care. How does the health care industry implement this decision?

At the opposite end of this flow, individual doctors treat individual patients with specific highly specialized care based upon very high complexity choices. Their decisions are based upon years of training and experience. The costs of individual treatments range very widely from tens of dollars to millions of dollars. The increase by 8% (so much and no more) must lead to changes in the decisions individual doctors make regarding the care of individual patients. They must decide what amount of time and attention to devote to a particular patient, which tests and treatments to perform or not to perform.

These decisions must be based upon trade-offs in health and care that compare diverse treatments. Physicians faced with restrictions on expensive procedures and treatments, or incentives to lower their own expenses, would have to make judgments about whether the amount of time and effort devoted to a particular appointment or individual, or a particular diagnostic test or therapy is "worth it," where "worth it" refers not only to the likelihood of a successful outcome but also to the cost-effectiveness of the decision to pursue it. Since this kind of judgment includes considerable uncertainties and it is largely incompatible with their training to treat disease, different organizations—and individual physicians—would make this judgment in different ways, resulting in extremely unstable and variable quality of care overall.

What can those who want to control costs do? It is clearly impossible for those who "manage care" to make decisions about care changes on an individual by individual basis in a way that will together correspond to the change in total flow specified from year to year. The only thing they can do is stipulate overall policies that act across the board. These policies typically restrict the set of options that are available for patients or physicians. Patients are restricted to certain physicians, hospitals or other care providers. Physicians are restricted in what diagnostic tests or medications they can provide. The amount of time spent in hospitals might be limited, or incentives to reduce the amount of time or attention to individual cases may be implemented. Would it surprise you to realize that limiting the options that a patient or physician can choose will have a negative impact on the quality of care that can be provided? Using across the board rules to control a highly complex system that is making careful decisions is not a good idea.

In the late 1960s, when health care costs started to rise well above the rate of inflation, and then skyrocketed, the need to control costs became clear. "Managed care" took hold widely in the 1990s after appearing in the 1970s and growing in the 1980s. Early on, managed care was conceived of as a way of providing more comprehensive high quality care. Today, it is mostly serving as a means of imposing cost-containment strategies designed to reduce overall health care costs. As we have discussed before, centralized hierarchical control and management will fail when the task is complex. Health care is a clear example of where this is happening.

People are increasingly aware of and frustrated with health care decisions being made by those who are concerned about the financial aspects of the health care system as opposed to their doctors, who are concerned about the individual. The incompatibility of these perspectives is clear. Efficiency methods are well suited to a mass production approach, but they are incompatible with the complexity of individual medical care.

Problems with cost control methods

This discussion clarifies why recent efforts to increase efficiency have led to organizational turbulence and the current need for and difficulties with quality improvement. As the treatment that is needed by individual patients has become more complex and individualized, health management organizations (HMOs), and other health insurance solutions have been striving to make its financial structure more large-scale and undifferentiated.

All the methods, which have been attempted to lower overall health expenditures at a national level over the last fifty years, have two things in common. First of all, they've largely been industrial-era efficiency methods. Second of all, by and large they've all failed.[37] Whether we're talking about the Nixon administration's wage and price controls in the early 1970s, or managed care's attempts to restrict medical treatments through drug formularies or limits on diagnostic tests, these cost control measures have at most produced a short-lived dip in spending before costs increased again.

The decisions made for these cost control strategies are overwhelmingly based on reducing large-scale sums that affect large numbers of cases rather than individuals. Due to the complexity of the problem of allocating financial resources, unexpected "indirect" effects have resulted from these efficiency methods. Indirect effects often impact on the quality of care that can be provided by physicians and hospitals. Moreover, the more problems arise with quality, the greater are the efforts to regulate the actions of doctors. Uniform regulation, whether for cost containment or for quality, has the same effect on a system performing high complexity tasks—diminishing overall effectiveness.

When people analyze what to do in order to improve quality in a particular case or type of case, the changes they decide to impose on the system will have impact on how care will happen for many other cases. Even if it helps for the specific case that was studied, when many different things are going on, the likelihood of causing more harm than good is very high. For a really complex system like health care, imposing uniformity is guaranteed to hurt more than it helps almost all of the time.

Because the resulting problems show up as indirect effects, the origins of the problem are difficult to uncover, and people who are involved generally don't recognize the relationship between the implemented policies and the negative outcomes. This makes it difficult to improve. New steps that are taken only move us father away from where we want to be. Frustration grows, improvement efforts become more and more essential, but continue to create more and more problems.

It is not surprising that the institutions that serve as intermediaries between the insurers and the doctors—the managed care industry, hospitals, and health care provider networks—have been undergoing dramatic changes in management structure and in patterns of delivery of care, and that every change may increase rather than alleviate the difficulties and turmoil in the overall system.

The problem is that the health care system is expected to behave efficiently with respect to financial flows at the large scale, but to exhibit high complexity of individual patient care at the fine scale. If all patients were in roughly the same condition, requiring roughly the same treatment, an efficiency approach would be fine. Streamlining works well for low-complexity procedures. However, for the high complexity medical treatment of patients one-size-fits-all does not work. Applying such methods can only result in poor quality care. To return to a theme from our discussion of warfare, you can't expect a tank division to move nimbly through a complex environment. This might appear to indicate that cost control is not possible if high quality care is to be provided, a difficult paradox to be in. Fortunately, although the above discussion of the current state of the health care system is grim, a fundamental approach to a solution does exist.

Large-scale health care

The resolution to this problem comes from recognizing that there are aspects of health care that *can* be treated with highly efficient processes. To apply methods of efficiency in the health care system, the first step is to identify which aspects of the system are repetitive and large-scale. Applying efficiency to those aspects makes sense and can save money. Applying them to the highly complex aspects, however, is not a good idea. Efficiencies in the system can be implemented in many ways if this distinction is carefully made. Here we will focus on the largest-scale parts of the health care system, those that should be dealt with at a population level. Indeed, although medical care and the treatment of disease are typically fine-scale problems, requiring complex individual attention through patient-physician interaction, these are not the only tasks that the health care system carries out. Which health services lend themselves to a large-scale efficient approach?

The answer is generally found in preventative care and public health. The aspects of health care that can be treated in the most efficient way include: wellness services, such as nutrition programs, management of some widespread chronic problems, prenatal care, and the treatment of common minor health complaints (allergies, stress, the common cold), and preventative procedures, such as inoculations and screening through diagnostic tests. Many of these services do not require individual decision-making by an independent complex agent (physician or other trained practitioner). They can be separated from those aspects of health care that require detailed decision-making and can be carried out using a population-based approach rather than through traditional one-to-one appointments.

It is worth recalling the history of today's health management organizations. The successful precursors of modern HMOs were designed with the goal of providing comprehensive health care to populations for whom such services were usually inaccessible. While cost-effectiveness was always crucial, the focus was on the improvement of medical care quality and members were subject to relatively few exclusions or limits. There was a major emphasis on preventative care and other services usually not covered by traditional insurance plans. The focus of managed care today is not the same. The central difficulty is that they are trying to perform all the tasks of health care using the same organizational structure.

Efficient health care/complex medical care

To solve the problems of the health care system, we argue that it is important to form two very different systems: an efficient system to deal with health issues that affect entire populations (and that can be made efficient on a large scale) and a system to address the complexities of individual medical care in an effective and error-free way. By separating simple, large scale "health care" from complex, individualized "medical care," we relieve physicians of tasks that can be addressed with a much higher efficiency, enabling them to focus their attention on the complex tasks for which they are uniquely trained. Not only does this create a more cost-effective health care system, it allows for a more effective and error-free medical system.

The high efficiency health care system depicted in Figure 10.3 would function in some ways analogously to a traditional mass production factory model. Some features of this system may seem disconcerting: it should be

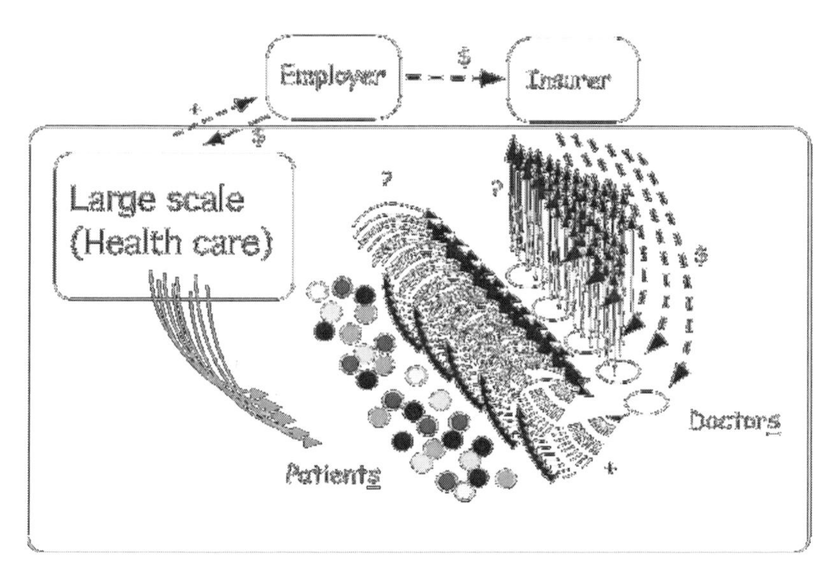

Figure 10.3: A proposed structure for a new health care system. One part is designed for efficient, population based health and wellness programs; the other part is designed for complex individualized medical care.

largely impersonal, not appointment-based, and not doctor-based. Nurses, technicians, and other non-physician practitioners can administer regular vaccinations and carry out routine diagnostic tests on large groups of people rather than through individual appointments. The purpose of the diagnostic tests is to ensure a high level of health in the population and to identify those who will need individual medical attention with a physician. The large scale system will not handle exceptions; all individuals requiring exceptional attention would be referred to the medical system. Most of the people using the large scale system would be well, and purpose of care is preventative. The objective of the well-patients program will be large-scale efficiency, but once a problem is identified, medical care for the sick can be highly personal and effective.

For example, a company could institute a mobile screening program, in which test equipment is brought to a workplace by the health care organization at regular intervals. Tests would be administered by technicians and results used solely for referral to a physician. An individual whose tests indicate that further actions must be taken would be advised to make an appointment with a physician in the medical system. The treatment of the individual may then require detailed and careful decisions performed by a highly trained team of physicians and other practitioners.

Employers, social organizations, community centers, and in some cases, government agencies including the Centers for Disease Control and Prevention and the Centers for Medicare and Medicaid Services, are the organizations that are naturally suited to caring about the population based health care. It might seem surprising to some but good employers care, even more than individual employees, about the health of their employees. Individual health is a key to productivity of the organization. Each individual has a small chance of being sick at any one time. However, a reduction in this probability can have a major impact on an employer. This means that employers and government agencies may be motivated to develop, and should welcome services from organizations providing population based care that will provide them at reasonable cost.

Screening/early detection: medical and financial effectiveness

An efficient health care system addressing population based care depends to a great extent on the development of effective screening and testing; and there has been much debate regarding the effectiveness of such techniques. Some of the concerns are medically related, while others are about their financial effectiveness. It should be recognized, however, that the knowledge of how to detect medical problems and perform early treatment is being developed and will increase rapidly. Moreover, a key aspect of the financial benefit from early detection arises from the large scale and efficient application of such tests. The existing system cannot carry out these tests efficiently on large numbers of patients because it is simply not set up to do so; and this is one of the main reasons why their financial effectiveness is under question. Before we can properly evaluate which tests will be effective when applied broadly, we need to change two of our basic assumptions: 1) that the tests will be administered by the existing appointment-based medical system and 2) that technology doesn't change. Some early detection tests that have been controversial are becoming more widespread in their usage, including mammograms and various other kinds of imaging including "full body scans." More traditional screening tests that are not widely used include the stress test for susceptibility to heart attack. These and other tests, if applied widely and systematically, can help to predict the level and type of medical intervention needed to avoid a medical disaster, without having to wait for more overt symptoms to occur. When frequent screening is done, it is possible to intervene when the time is right, as opposed to responding with urgency to the first indication of symptoms.

Not all tests are a good idea. Still, to develop a perspective on evaluating when tests are constructive, it is helpful to compare the introduction of these tests with the introduction of new technologies in other industries, for example, the consumer electronics industry. We are now seeing the introduction of high-definition television. If we studied this technology a year or two ago, we would find that it was not cost effective and not broadly useful. The way it was introduced, however, was by starting with high cost versions that only a few people could afford. Then gradually, as both the technology improved and the volume of production increased, it became accessible to many people and financially viable for the companies that are producing it. How did the companies know that this would work? First, they didn't know for sure. Still, they had experience with previous generations of consumer electronics. This experience told them that technology improves with time, and as adoption increases, mass production reduces costs. When we think about health care we don't think in the same way because the system is not designed around mass production and scientific medical studies are not allowed to suppose that we might learn more in the future about how to use the information that we gain from medical tests.

Highly efficient, rapid, and cost effective performance of tests and inoculations will lead to improved efficiency and relieve the financial pressure on the medical treatment of individual patients. There is another industrial example that provides a useful analogy. There have been studies and changes in practice in preventative care and equipment maintenance in factories that have had dramatic effects.[38] Preventative maintenance does not reduce costs immediately. Initially, there is a great deal of work to be done because problems are detected earlier and much work must be done to repair the broken equipment. However, this eventually leads to lowered overall costs, as the reduced failure rates from properly maintained equipment reduce the failure rates later on. On the other hand, poor maintenance catches the system in a vicious cycle of failed equipment and overtaxed maintenance crews performing interventions in a crisis context. Studies show that this later case is where you spend more and get less in terms of equipment reliability! It is not too hard to see the analogy between this and the current situation in health care, where we are spending more and getting less from our health care system.[39] Many countries using other health care systems focus more attention on public health than the U.S. This does not mean that they have the balance right (even more public health might be better, or more individualized care might be needed), but it suggests that we are moving in the wrong direction when we focus on cost containment and efficiency in the treatment of individual patients. Implementing preventative tests and early diagnostic techniques will initially require a greater investment, but with application of such tests on a large scale, a significant and permanent decrease in costs should follow. Better yet, we can spend the same amount of money and achieve a much higher quality of life through improved health.

The underlying message of studies of equipment maintenance is simple and clear. However, reaching the point where organizations behave this way is not necessarily easy. Quite generally, a short-term perspective of treating just the problems that you see at the moment is ineffective over the long term. Still, starting to take the long-term view will make matters worse (at least in cost and effort) in the short-term. The overall key to success is developing a long-range perspective and sticking with it! This perspective is easier to establish if it is possible to experiment in a local context, thus promoting wider application.[40]

Conclusion

What goes by the name "health care" right now is an individualized system. Despite the fact that many of its services are largely universal, population-oriented ones, the system provides these services mainly through the traditional one-to-one physician-patient model, so that it can provide individualized medical care when such care is needed. The problem is that this one system is expected to provide both financially efficient health care and complex medical care, therefore it should not be surprising that it is struggling with this dichotomy. Efforts to lower costs through managed care and other insurance and care delivery schemes must lead to ineffectiveness, which is manifested in medical errors and decreasing quality of care. A fundamental solution requires separation of complex tasks from large-scale tasks. Individualized care should be entrusted to a fine-scale medical system, while a distinct system should be created for large-scale and efficient heath care or wellness programs. The large-scale financial structure that currently drives the health care system will then be matched to an efficient, population-based care delivery system, relieving much of the turbulence caused by the current allocation problem. The result: a healthier population, a more focused culture of high-quality medical treatment, a relieving of pressure on our overtaxed medical practitioners, and, perhaps surprisingly, lower costs.

Improving any organization's performance involves assessing and dealing appropriately with two aspects of the system's capabilities and tasks: scale and complexity. We must recognize and distinguish between scale and complexity and approach each accordingly. Efficiency can be applied to problems involving scale—the repetition of many identical actions—but not to problems involving complexity. If you try to make a simple and large-scale process more complex, you're simply wasting your resources. You can lower costs by making large-scale processes more efficient, but if you try to lower costs through efficiency in a complex process, you'll end up with errors. Apply efficiency for large scale tasks, and complexity for complex tasks, these are the key lessons.

Chapter 11

Health Care II:

Medical Errors[41]

Introduction

In recent years, the health care industry has grappled with an increasing awareness of its own fallibility. An Institute of Medicine (IOM) report[42] released in 2000 announced that preventable medical errors of all kinds are killing between 44,000 and 98,000 people per year—more than the number of deaths due to automobile accidents or breast cancer. While the methods of counting and accounting for medical errors are disputed, the problem has become acute no matter whose numbers you trust. From the first part of the book, we know that the existence of many errors generally implies the system performing the task is not complex enough. Medicine and medical practice are incredibly complex and the existing system is just not equipped to deal with this degree of complexity.

The dangers associated with receiving medical care have become a growing concern for the American public. Dramatic examples of medical errors—often fatal—appear regularly on the front pages of newspapers and the covers of magazines. The public's perception of medical errors is often dominated by a scapegoat mentality that prompts reporters and readers to assign unambiguous blame to a particular individual, procedure, or device.

The need for widespread improvement has been recognized, but it is not always clear what kind of framework can help health care providers understand how these errors come about. The IOM has emphasized that the key to reducing medical errors is an understanding that they are "systems related" and not attributable to individual negligence. Recognizing that the errors come from system design is a good first step, but it doesn't actually tell you how to improve the system to prevent the errors from occurring. In this chapter we will look at the properties of systems that can provide effective and error-free medical care. Such a medical system includes effective individual practitioners, but also includes effective communication channels and effective coordinated behaviors of multiple individuals. Explaining the properties of a system that can perform complex medical tasks turns out to be only part of the picture. A complete discussion, requires an additional step to understand how we can create the effective system and continue to improve it over time, a topic that will be taken up in Chapter 15.

Prescriptions and the problem of providing medication to patients

There are many aspects of patient care where medical errors arise. We will discuss the example of the prescription error problem, one of the most common and extensively studied forms of error. The lessons we will learn from this example can also be applied to other areas where problems arise.

Providing medication is an important—and complex—service that the medical care industry provides. One way of understanding the complexity of a task is to count the number of possible options. How complex is the task of drug prescription and delivery? Today there are about 15,000 registered drug names in the United States. Supplying the right medication for a patient, then, means making sure that he/she receives the right one of those 15,000 possibilities, but that's not all; not only are there numerous drugs available, there are many possible dosages—both quantity and timing—and methods of administering them. With all of these different parameters, imagine all of the possibilities, many of them potentially harmful. Nurses recognize this complexity and use a five "rights" mantra —the right patient, the right drug, the right time, the right dose, and the right route. Given a high-complexity task, where there are many wrong outcomes for each right outcome, errors are likely to occur. Conversely, if many errors are taking place, it's very likely that there's a high complexity task that isn't being dealt with effectively by the existing system. The problem with the system for providing medication is that for many years it hadn't been revised to accommodate the increased complexity of its task. Today there are many efforts to improve the system, however, to make new systems work well, it is important to understand why the old system is failing.

For example, let's imagine how the traditional system might work for inpatient medications. The doctor writes (or scribbles) the prescription on the patient's medical chart, possibly using certain well-established abbreviations. Then, a hospital employee copies it from the chart. The copy is taken to the pharmacy, where a pharmacist reads and fills the prescription. He gives the medication to a hospital employee (perhaps the same one, perhaps not), who then transports it to the appropriate area of the hospital, where a nurse administers the medication to the patient.

Let's start by examining one segment of the process: the doctor writing the prescription on a piece of paper. Theoretically, a doctor has 15,000 possible medications to choose from when writing a prescription. One high-profile aspect of this proliferation of choices is name confusion. Take the example of these two drugs: Celebrex and Cerebyx. Celebrex is a prescription medication that provides pain relief from arthritis. Cerebyx, on the other hand, is an anticonvulsant prescribed for the treatment of seizures. Name confusion has led to mistreatment of patients and this pair is only one example of the many pairs of similarly named drugs that have caused confusion in the writing or filling of prescriptions. Some further examples are Lamictal (an anticonvulsant used to treat bipolar disorder) and Lamisil (an antifungal drug); Zyrtec (an antihistamine) and Zantac (an ulcer drug); Sarafem (an antidepressant) and Serophene (a fertility drug).

Prescription errors also occur in specifying the correct dosage. For example, in a highly publicized case in Washington, DC,[43] a surgeon wrote a prescription for ".5 milligrams" (not "0.5") of morphine for a nine-month-old baby, to be administered by a nurse after a series of operations. The unit clerk transcribed this number as "5 milligrams," without a zero or a decimal point and the medication was dispensed in that amount. The nurse tending the child followed the order, and due to the erroneous dosage—ten times the intended amount—the child died.

Given this account of what happened, we might blame any one of the people involved in the prescription-filling process. We could argue that the doctor made the crucial error in leaving off the extra "0" before the decimal, making the number more open to possible misinterpretation. Or we might insist that the clerk's misreading and misfiling of the prescription was responsible. We could also contend that the nurse who administered the medication should have recognized that the dosage was too high for a small child. In the flurry of attention that followed this case, all of these hypotheses for who was "at fault" were proposed.

Space of possibilities

This account of the events leading up to the tragic error, however, leaves out the most important information of all: the space of possibilities for each step. Each of the individuals involved in this case had a distinct set of possible choices in the actions that he or she took. The set of possibilities for each task determined the likelihood for error. Without understanding the space of possibilities, we simply cannot evaluate the system to determine where the errors are coming from.

For example, what if morphine were only administered in the amount of 0.5 milligrams, to any kind of patient? If this were the case, the pharmacist and the nurse should have known that there's never, ever an instance in which 5 milligrams of the drug should be dispensed. On the other hand, if morphine were usually administered at 5 milligrams, then more respon-

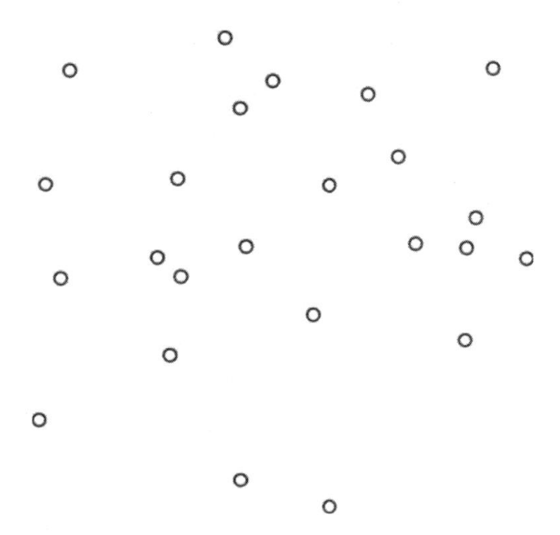

Figure 11.1: The space of possibilities: Each dot represents a possible valid decision.

sibility might lie with the doctor, who should have been more careful to emphasize that this was an exceptional case by adding the extra zero and perhaps making the decimal point more visible.

Figure 11.1 presents a graphical illustration of the space of possibilities for this problem. The dots represent the set of possible outcomes for the decision-making process in administering medication to a patient. Each dot is a possibility that under some circumstances could be correct; each possibility is defined by the type of medication, dosage, route, patient, and time of administration. For a given situation, we want one and only one of these possibilities to occur.

Ideally, when a doctor writes a prescription, he will record information that corresponds to a complete description of one of these possibilities. Then, through the process of filling the prescription, the correct choice should be made. Now, loosely speaking, the complexity of a system is the amount of information needed to determine which of these dots is the one that has happened (or should happen). One measure of this is the length of that description—the number of letters, perhaps, used to record it. Therefore, the complexity of a particular prescription can be measured by the length of the description the doctor has written down.

What happens when errors occur? If a doctor miswrites a letter, or a pharmacist misreads a letter, the prescription no longer describes exactly the correct possibility. In Figure 11.2 the rings around the dots represent

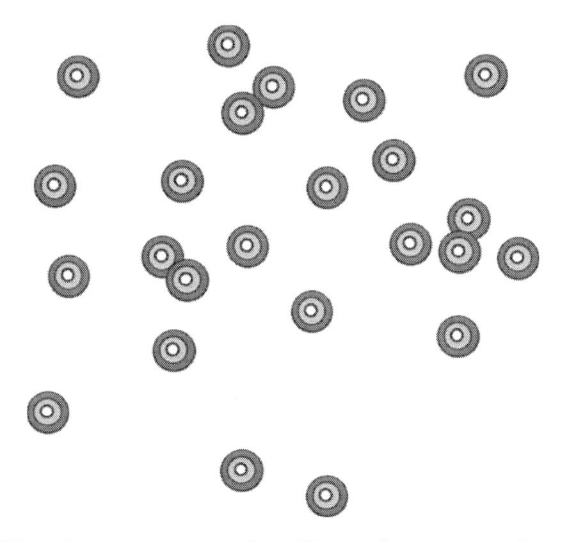

Figure 11.2: The rings represent the effects of errors. As long as the rings don't overlap the intended (right) decision can be inferred.

these errors of perception. Accidentally switching one letter for another in writing the name of the medication, for example, would correspond to moving out to the first ring around the dot. Another error would place us in the second ring. Because some of the error distances around the dots overlap in this space of possibilities, a small number of errors can take us from one distinct possibility to another.

Suppose a doctor were writing a prescription for Cerebyx, but he makes three errors in his handwriting: he accidentally writes an "l" instead of "r," and an "e" instead of "y," and then inserts an "r" after the "b." With those three errors he's now written "Celebrex," a completely different drug. His errors have moved him from specifying one possibility to specifying another. If he only makes one of those errors (and writes "Celebyx," for example), the prescription will lie somewhere between possibilities. At this point, it's ambiguous which possibility is called for—did the doctor mean to write "Celebrex" or "Cerebyx"?

From this discussion, we can see that it's crucial to understand the structure of the space of possibilities. If there were no drug with a name very close to Cerebyx, then one or two wrong letters might not make such a difference. (For example, if there were only two drugs available on the entire market, Cerebyx and Prozac, then accidentally writing "Celebyx" would still unambiguously point to Cerebyx.) If all medications were administered to patients at a dosage of 0.5 units, then even the dropping of the zero from the prescription would not lead us ambiguously close to any other possibility—because there would be no other possibilities for dosage. The further away the dots are—or the fewer dots there are at all—the less likely you are to make enough errors to cross the space between them.

Error correction

In response to the medical error problem, many organizations have produced recommendations, proposing a variety of procedural, organizational, and technological changes that hospitals, clinics, and pharmacies can carry out to reduce errors. Many of these recommendations are quite reasonable. However, unless you have a good understanding of the system you're trying to change, it's difficult to understand which changes will really help. It is also often hard to motivate people to make a change without being able to explain why it should work and even how well it should work. The key is understanding the fundamental role of complexity and scale and this is what we will use to analyze the proposed changes.

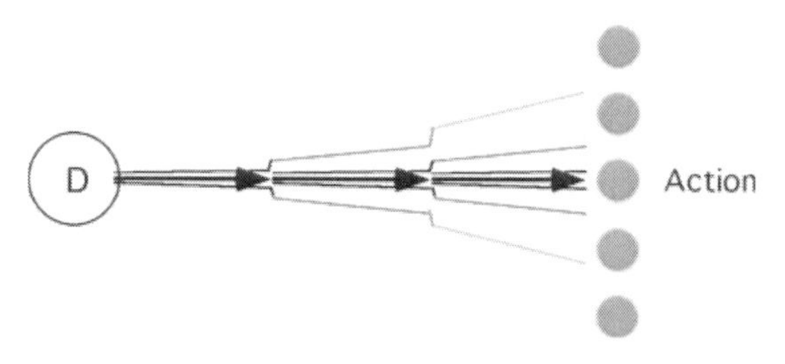

Figure 11.3: A decision followed by several communication steps, with possible errors, and then action.

There are five quite natural approaches to reducing medical errors: feedback correction, eliminating steps, redundancy, automation, and reducing the local complexity of the task. The first four when appropriately used and effectively implemented can ensure that a decision that has been made is actually carried out. The last (reduction of local complexity) also has another use: reducing the likelihood of decision-making errors. Each of these approaches will be discussed below.

In talking about these methods of error reduction, it's important to be clear about one thing: we're not talking about subtle errors of judgment in the actual medical decision regarding what treatment is necessary. The errors we're talking about are obvious differences between what should be done (as decided by the physician) and what is actually done.

The diagram in Figure 11.3 illustrates the process we are concerned about. One decision maker, D (usually a doctor seeing a patient), makes the decision about the right action to perform. This decision is communicated through a series of intermediaries who carry out the intended treatment. The methods of error correction address the problem of deviations from the desired track—resulting in a different treatment than was intended by the doctor.

The model of medical practice described in this diagram doesn't always apply, but it is helpful to think about this as a first step toward considering some of the key issues associated with medical errors. Still, we should be aware of the assumptions it requires. First of all, we assume that the only decision-maker in this process is the doctor. Under this assumption all other health professionals, be they nurses, technicians, or pharmacists, simply carry out the practical details of a decision which has already been made. The pharmacist simply translates the information from the doctor (the prescription) into the medication, and the nurse just administers that prescribed medication. The second and more subtle assumption is that the pharmacist receives only instructions from one kind of doctor (not one doctor but rather one kind of doctor).

Neither of these assumptions is necessarily true. Later in the chapter we'll discuss their limitations, and what happens when they don't match what's really going on. Starting with a simplified picture will enable us to introduce the basic strategies for preventing errors. When we add more real-world complications we can see how they affect the usefulness of each of these strategies.

Feedback correction—check once, check again

If errors are occurring in your system, one way to remove them is to put checking procedures in place to catch errors that have already occurred. For drug prescription and delivery, this kind of "feedback correction" involves double-checking the prescription at the end, or possibly at various stages in the process.

The most direct way would be for the doctor herself to check the medication before it's administered to the patient. Ideally, here's how this would work. The doctor writes a prescription in the hospital, which then follows the ordinary routes—it's taken by the hospital employee to the pharmacy, where the prescription is filled. Then, the prescription returns to the doctor, who checks to make sure that the medication is what she meant to prescribe in the first place. If so, then she gives the green light and everything is as it should be. If it's not, then the error has been caught and the prescription is sent through the same process again.

This scenario might seem sort of unrealistic. Doctors have a lot to do and it is unreasonable to expect that they will double-check every medical procedure in the hospital. Furthermore, coordinating the doctor's schedule to be at the right place and time would be ridiculously hard. However, there are more feasible ways to carry out this approach. For example, if the particular medication needed is accessible nearby, this kind of double-checking by the doctor might be possible. This happens in limited ways already for some medications that are "on-hand" on the hospital floor, emergency room or even doctor's office.

The existing hospital procedure also already has a more general double-checking procedure. The prescription written by the doctor stays on the patient's chart, which is kept near the patient (for example, at the nurses' station). A copy of the prescription is made, which is the copy that is taken to the pharmacist. Once the prescription is brought back to the patient, it can be checked against the original that remained with the patient. Of

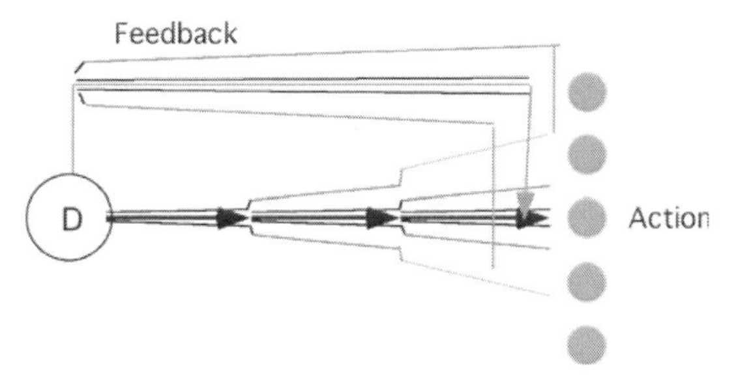

Figure 11.4: Adding a second path of information allows for feedback correction, but setting up and coordinating the extra path often creates its own opportunities for failure and surely is a lot more effort.

course, as we've found out from the case in Washington DC, the additional copying adds a step that might itself introduce errors. Making sure that the initial copy is a good one requires care and automatic methods like using carbon paper, a photocopier or a fax machine, may ironically add other opportunities for problems (poor copy quality, malfunctioning equipment, and a need for adequate supplies, repairs and backup systems).

With all double checking procedures, we create two paths for the information instead of just one (see Figure 11.4). One copy of the prescription is sent to the pharmacist, where it is filled. The other route doesn't directly involve the medicine; its only function is to keep an accurate record of the information in the prescription used to determine the medication. Once the medicine is obtained, it is double-checked against the other copy to make sure that the medication is the one that was originally prescribed. This type of double-checking procedure would catch errors that occur between the act of the doctor's writing the prescription and the actual administration of the medication.

However, there are several problems with this approach. First, this approach creates additional steps were errors can be introduced. Creating an extra path for information requires at least two additional acts, one at the beginning when the information splits into two paths, and one at the end when they are checked against each other. Moreover, this approach would not catch errors in the first step of the process, when the physician actually writes the prescription. If the prescription is written with an error and we duplicate the prescription, the error now exists in both copies. Because the first step of the process plays a special role, we will pay particular attention to it in the discussion of each of the strategies for reducing errors. To solve this problem, we must consider the act of writing out the prescription as

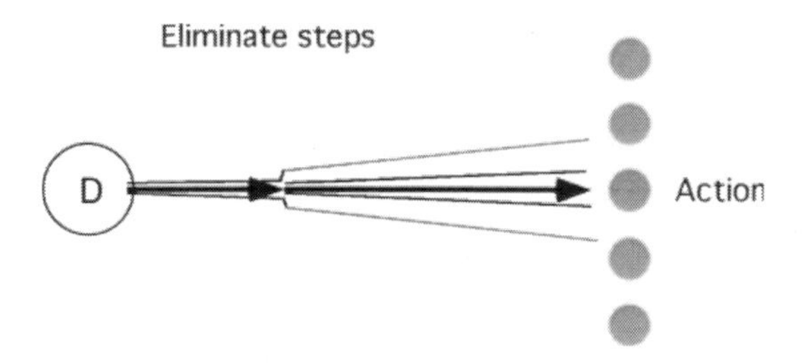

Figure 11.5: Removing unnecessary steps (here there is one less step than in Figure 11.3) reduces the possibility of error when those steps are likely to cause errors.

the first step in the communication channel and find some way to duplicate that step. Before we discuss how this can be done, however, let's consider the possibility of removing unnecessary steps.

Removing unnecessary steps

Another important approach to reducing errors is to eliminate steps that might introduce errors (Figure 11.5). In the Washington DC case, for example, if the prescription had not been copied, perhaps the decimal point would have been noticed and the patient would have been given the right prescription.

If a current procedure contains unnecessary steps, removing the unnecessary steps reduces the likelihood of errors in the original process, which is better than having to eliminate errors once they have been made. This is also a good approach for reducing the amount of time needed to complete a process. For example, recently in the emergency room of a different hospital in Washington DC, the number of steps required to receive the results of a blood test was reduced from 8 steps involving 7 people and taking about 60 minutes to 3 steps involving 3 people and taking only 3 minutes.[44] This change in procedure was achieved by placing a small blood-testing facility right in the middle of the emergency room. With this arrangement, the person who draws the blood can immediately take the sample to the testing location, rather than having to send the blood sample to a different part of the hospital, saving a great deal time and making the overall process much more efficient.

However, there are also some problems that may occur when we try to eliminate unnecessary steps. First, the extra steps may be needed for other purposes. For example, if we want to have feedback checks as discussed in the last section, then extra steps are necessary. While reducing the number of steps reduces the likelihood of error, if we eliminate key checks we may actually end up increasing the number of errors. Evaluating the trade-off (between adding steps that allow checks and removing them so they don't add more error) requires careful thought. Eliminating communication steps is also not possible when you want people to work together sequentially so that the task can be distributed among them. This often is the case when specialists or special equipment are necessary for part of the process.

Moreover, the approach of eliminating steps, like feedback checking, does not affect the very first step: the writing of the prescription. We can't eliminate that step (unless the doctor administers the medication) and preventing errors in the first step is important to ensuring that errors will not occur.

Redundancy

The third approach to preventing errors uses redundancy. To create redundancy, one starts with more information at the outset of a procedure. The key to reducing the chance of error is to obtain more information from the physician at the start of the process. This information then follows along the entire route of the process so that everybody on the way can check to make sure that what they are doing is correct, thereby reducing the overall likelihood of errors (Figure 11.6). Having a lot of extra information could be burdensome, but it turns out that even just a little more is enough to reduce errors dramatically.

To implement this approach for medications the doctor would include twice as much information about the desired medication on the prescription. Using more words than the minimum necessary to specify which possibility is intended, provides a redundancy that can help to eliminate errors. If we consider a prescription that would be implemented correctly

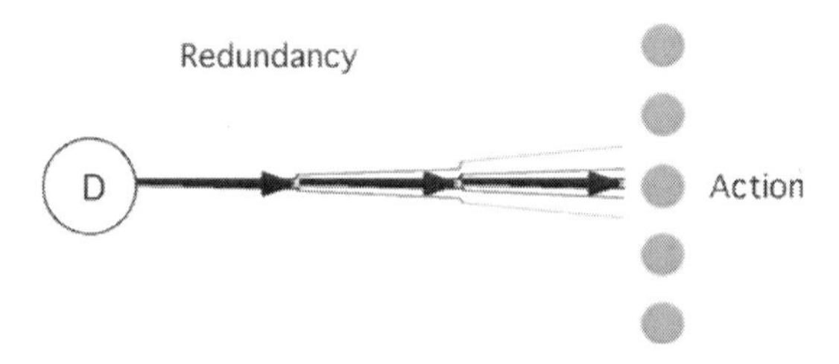

Figure 11.6: Redundancy adds more information, thus reducing the likelihood of errors at each step (including the first one), causing the spread of possibilities to decrease as shown.

without the extra information, adding additional words gives nothing new and seems unnecessary and a waste of time. However, when there are errors and the extra information leads to a correct choice, it makes all the difference in the world.

For example, all doctors could write down on prescriptions both the generic and trade names of a drug, or they might write the name of the drug and the condition that it's being prescribed for (the indication). Any kind of additional information that could be used to identify the drug needed could be required on the prescription—including, even, the shape or color of the packaging. This information is redundant, but it serves as a check on the other, standard form of description.

If you always write the drug name and the condition, then accidentally writing "Celebex; Seizures" instead of "Cerebyx; Seizures" would still indicate very clearly that you're prescribing the anticonvulsant Cerebyx, and not the pain medication Celebrex. By increasing the amount of information you're giving about the prescribed medication, you make the space of possibilities more and more dispersed. In effect this increases the distance between the dots in the space of possibilities shown in Figure 11.1, because the number of errors that would be necessary to go from one medication to another is quite large when there is more information. As the dots move farther and farther away from one another (with added redundancy of description), errors are less likely to matter. If the dots are farther away, then even four errors won't lead to any dangerous ambiguity.

This is also the reason why physicians are advised to write 0.5 and not just .5 when writing prescriptions. The former has enough information to be interpreted correctly most of the time while the latter is more prone to error because the redundancy is low.

Like feedback correction, adding redundancy to a procedure means adding time to the doctor's task. However, because of this redundancy, for example, having a prescription with both the name of the drug and the condition on it, the pharmacist (or the nurse, or the patient) may notice and be able to correct errors before administering the medication. In terms of complexity and scale, this is the same process as having the doctor re-approve the medication before it's administered. In both of these procedures, you're doubling the information that comes from the doctor so that the two sets of information can be checked against one another. With redundancy you double the information at once and the two sets of information, physically attached to each other, can be checked against each other at every step. With feedback the two sets of information are kept separate, and you check them against each other at a specific time later on. These two approaches are not exactly the same in the way they avoid errors, but they are close.

One of the crucial advantages to the redundancy approach is that it reduces the impact of errors in the very first step, the writing down of the prescription. No matter how the physician communicates the information the first time, including in automated ways that we will discuss next, the issue of making sure that this step is done well is crucial and redundancy can help.

Automation

Automation involves identifying processes and chains of events that don't require complex decisions and making them more efficient by introducing computers and communication technology. This often also reduces the number of people or steps involved to eliminate handoffs or communications that may cause error.

Why are computers helpful in reducing errors? To start with, it's because they are less complex than people. Introducing a human being into a process produces the potential for change, because human beings are so complex. People are better at making subtle complex decisions than they are at automatic (rote) execution of simple tasks. For a given situation, there are potentially thousands of possibilities for what a person might choose to do. A computer, on the other hand, is not nearly as complex as a human being. It carries out repetitive, simple logic very reliably.

Automation is one of the most reflexively suggested methods of error reduction, but it's not always the answer. It is interesting that computers are proposed as the best way to avoid human errors when the most commonly used computers frequently crash. Perhaps people have been watching too many science fiction movies! There are two key problems with automation: correct implementation and an effective user interface. If the system is not implemented correctly, the system will make many errors. This illustrated in Figure 11.7 by showing the process moving in the wrong direction. Since people believe that automation is the answer to solving problems, they will usually blame the programmer for implementation errors rather than the approach of using automation itself. If the user interface is not done correctly, there will be many errors that occur at the first step of the process, when the equipment is instructed what to do. When such an error happens people generally blame the person who entered the information rather than the user interface and do not think of

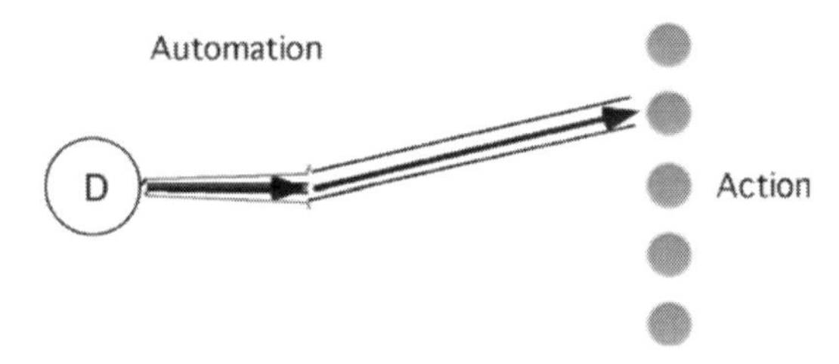

Figure 11.7: Automation reduces the likelihood of random mistakes occurring in the steps that have been automated, but may introduce problems at the starting point and through hard to find errors in implementation (bugs).

blaming the strategy of using automation itself. As the complexity of tasks increases, it becomes increasingly difficult to make sure these two are done correctly. Indeed, sometimes it is much more difficult than having people do them correctly.

More generally, to really make proper use of automation, it's crucial to understand what it is useful for. Automation will make things more efficient when the execution of a task can be uniquely and fully specified without any further decisions being made. It can also reduce error due to the elimination of unnecessary intermediate steps. For example, if, as we've been assuming, the doctor's prescription uniquely specifies what should be administered to the patient, then the process of fulfilling the doctor's decision involves no further decisions after the prescription is produced and automation can help. A hospital might set up a direct channel of information from the doctor to the pharmacy. The doctor would fill out an electronic prescription entry form that would immediately be sent directly to the pharmacy and printed out. The pharmacist would also do no copying whatsoever and would simply fill the prescription as printed out. We might go even further and install automated dispensing units, at least for common drugs, that receive the prescription and dispense the medication without any human intervention at all. Automated dispensing systems are already being implemented in some places.

Let's look more carefully at the first step, the doctor writing the prescription. One part of this step in which automation may seem to be helpful but is only if it is done correctly, is in providing an additional immediate feedback check at the time of writing the prescription. A physician generally checks the prescription immediately after writing it. She reads it to check that it is clearly written as far as she is concerned. An electronic entry system can be designed to enhance this immediate check by having a typeface version of the written prescription or by automatically showing additional information like the standard medical indication for that prescription. While this may be helpful, some words of caution are worthwhile. This automated process seems like the same as the case where the physician wrote additional information. It isn't. The additional information is not coming from the physician, it is only being verified. Verification has a lower level of reliability because it requires much less information from the physician. A physician is less likely to misspell a prescription and write the wrong indication (that happens also to correspond to the same drug that the spelling mistake gives), than to "blindly" approve an incorrect indication suggested by the electronic entry system. Thus, even if automation is used, it is better to have the entry system require the physician to enter both the medication and the condition. The key is to realize that the process of information transfer from the physician to the communication channel should not be made efficient. Despite the great desire to make it easier, the key to avoiding errors is to require more information from the physician as opposed to less. Once electronic entry is completed, feedback checking at the time of administration will be easier. Feedback involves sending information into two channels that contain identical data, which can then be checked against each other. One set of information is transferred to the pharmacist and translated into the medication, which then physically passes to the patient. The other channel is the feedback channel, which will simply contain an electronic version of the doctor's original prescription. Since the electronic version can be sent around automatically in any number of copies, the feedback process is simpler and, if the equipment is reliable, more reliable.

Feedback checking could be further augmented by having the computer read the package and do the comparison of medication with prescription (rather than the person administering). A few pharmacies and hospitals have adopted barcoded drug selection procedures, in which a paper prescription includes a computer-generated barcode that can be deciphered automatically at the pharmacy before dispensing, and even at the patient's bedside directly before administration. According to the FDA's recent regulations (February 2004) most prescription drugs, and over-the-counter drugs frequently used in hospitals, will be required to bear a bar code uniquely identifying the drug, its strength, and its dosage form. Checking this information with a barcode reader at the patient bedside—especially in conjunction with barcoded patient bracelets—could catch errors involving the wrong medication, dosage, timing, or patient.

There are many other useful ways for introducing automation into a system—electronic medical records, hand-held wireless computers for bedside use—but recognizing what they can improve and what they might not be able to do is important. Electronic medical records are important in enabling easy retrieval or sharing of information. However, among other issues, ensuring that the most important information is brought to the attention of the person who needs it, is not easy to guarantee. Hand-held computers can help in various tasks including checking medications. Some of the potential for improving the system and the required care in execution is described above. In each case the choice of what to automate and the quality of implementation are crucial to the development of effective systems. Since the existing systems have been developed and refined over many years, it will be difficult for new systems to introduce improvement unless great care is taken.

Two decision makers

Up until now we have been assuming that there is only one decision maker in the system—the doctor. However, it's not that simple. Pharmacists make decisions—they're not just following instructions. Pharmacists are often responsible for determining whether multiple drugs prescribed to the same patient are incompatible, that is identifying harmful drug interactions before they occur, or making substitutions of one drug for another.

How does the complexity of the pharmacist's decision-making affect the possible solutions we've discussed so far? Some of the suggested improvements may not work as well or even at all, while others survive unscathed.

Feedback correction now has a problem. Since the pharmacist can make drug substitutions, there may be good reasons that the prescription is not the same as the drugs that are administered. A simple feedback checking process will not work. The checking process, whether manual or automated, has to be able to figure out whether a substitution is OK or the result of an error. Either the person who is doing the checking or an automated system that performs checking must recognize which substitutions are reasonable and which are not.

Adding redundancy in the prescription still works. It improves the communication channel to the pharmacist without interfering with the pharmacist's decision making and allows him to modify the prescription if appropriate. In this case, writing both the medication and the indication seems like a really good solution.

Eliminating intermediate steps between the physician and the pharmacist, or between the pharmacist and the act of administration may still be helpful, but it cannot eliminate the pharmacist involvement. The same is true of automating steps in the process. The automation should not interfere with the decisions that are made by the pharmacist. Indeed, understanding the decision making role of the pharmacist is a key issue in determining whether or what steps to automate.[45]

The many-to-one communication channel problem

Let's expand our view of the system one more step to observe that there are many different physicians sending prescriptions to the same pharmacist. Because of specialization there are many different types of physicians, and each specialty will tend to have its own set of most commonly prescribed drugs. While there are some drugs that a neurologist (a specialist in nervous system disorders) and a rheumatologist (a specialist in arthritis) might both prescribe, there are many others that are particular to each specialty.

From the neurologist's perspective, there's little to worry about when writing a routine prescription for Cerebyx for an epileptic patient. If there were another drug with a similar name that the neurologist tended to prescribe, he might naturally be more careful to identify clearly which he meant. However, neurologists don't often prescribe Celebrex and from his point of view the communication path to the pharmacist may seem good enough. Similarly, a rheumatologist whose prescriptions happen to pass through the same pharmacist may see little likelihood for confusion in her own writing of a Celebrex prescription.

This is not at all how the pharmacist sees it! The problem is that while the physicians have no need to think of both possibilities, the pharmacist is faced with both regularly and confusion is very likely indeed. More generally, there are many more possibilities on the pharmacist's side of the communication than on the physician's side. Having many different people communicating with one person places a very high demand on the communication channel at the far end. This is why it is not really enough for the physician to consider his own handwriting and ask himself if it is clear enough. He must consider what the pharmacist sees—how many possibilities the pharmacist has to distinguish among—to appreciate what he really needs to be clear about.

If one of the major problems lies with an unbalanced communication channel why hasn't this view received more attention? The answer is quite simple: differences in perceived authority between physicians and pharmacists. Physicians are assumed to be more important and to have more power than pharmacists. Because of this, even if a pharmacist is uncertain about what drug the physician is prescribing, he might be reluctant to call up the physician to double check. Power is a key aspect of how roles are designed in an organization, and the weak points in organizational effectiveness are often determined by how power is perceived.

Differences in power generally are a way of shifting burdens from the powerful to the less powerful. Among these burdens is that of complexity. When physicians are powerful they can shift some of the complexity of their tasks onto others. If pharmacists become more powerful they could shift some of their complexity back to the physicians. This might not be a good idea if physicians have to respond to many calls from pharmacists at a time when they are already overburdened. Understanding which is better can only come from a more careful understanding of how the complexity of tasks is distributed through the system.

Now that we understand the problem and why it exists, how can we change the system to address the problem? Insisting on the physicians writing the prescription in two ways, for example, the drug and the condition (indication), seems like the most direct solution. Alternatively, instead of the condition, the physician could write his or her specialty. This would be a weaker, but possibly sufficient way to include extra information. Marking the physician specialty might be done in a partially automatic way, by using an electronic identification system. This way the system can distinguish between what different physicians may write, without imposing special rules to make it work.[46]

The next step in our discussion will change its focus. While we will still consider the communication channel, we can also use the next approach to address wider issues in the organization. This is important because the communication channel itself is not the only reason that errors occur. The wider view is necessary to address many other sources of error and thus to solve systems-related medical error problems.

Reducing local complexity

The use of feedback, elimination of steps, redundancy and automation can help with the problems associated with communication channels. They reduce the impact of errors in, increase the capacity of, or reduce the error rate in the communication channel. However, in many cases the source of errors may be the complexity of the tasks the individuals within medical system have to perform. The number of possibilities that medical practitioners face at every decision may be too large. A crucial method of error

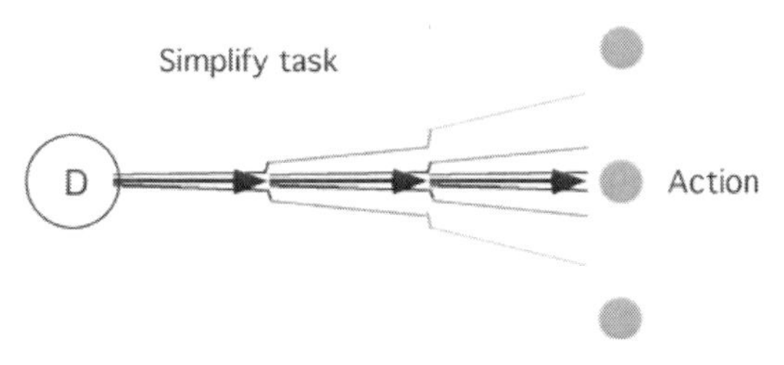

Figure 11.8: Simplifying the task reduces errors by making fewer options, shown here by reducing the number of dots representing valid actions. If fewer options are needed to perform the task, then this can work. Otherwise, specialization has to be part of this approach.

correction is to reduce the number of options available at any given step in the process (Figure 11.8). By limiting the set of possibilities that could be chosen, you reduce local complexity, reducing the demand on the system, and thus decrease the likelihood of mistakes.

There are two ways to reduce the complexity that a person has to deal with. The first is to simply reduce the number of actions that the entire system can execute, and thus the number of possibilities the individual has to deal with. The second is to divide up the many possibilities among multiple individuals. Whatever changes are made to reduce local complexity, it's important to assess whether the overall task still has sufficient complexity to be effective. This is the crux of the problem of organizational effectiveness: you want your system to perform high-complexity tasks, but with individual local tasks that are simple enough that errors are unlikely to occur.

Reducing unnecessary possibilities: standardization

The elimination of possibilities starts from the recognition that in practice, you don't always need *all* possibilities that might in principle be used. We see this process of stripping away unnecessary possibilities in many forms of standardization. For example, in the past pharmacists were responsible for mixing ingredients to produce medications in various forms (liquid solutions, ointments, powders, tablets and capsules). Today, however, pharmacists do much less mixing and packaging of the drugs, which usually come prepackaged in standard forms. Also, nowadays the dosage for many drugs is the same for all adults, often administered at fixed times twice a day. Drug packaging provides only a few options for how it can be administered. All of these changes reduce the set of possibilities tremendously. As long as the possibilities at your disposal correspond to the possibilities needed for treatment, this reduction in complexity is a very good idea.

In general, standards of practice lead to reduction of complexity. To the extent that we can be certain that the possibilities we are eliminating are absolutely unnecessary, this is great. However, when people develop standards they often consider only the "typical" or "average" case and create standards that do not apply to the space of all possible cases. Even in the case of standard adult drug doses there is the potential for problems: the same dose can have a very different effect depending on whether it's administered to a football player or a jockey. This is the danger inherent in standardization: reducing complexity when it is needed for effective action.

Automation provides additional methods for standardization and constraints on the possibilities. For example, an automated system of electronic prescriptions could be used in quite reasonable ways to constrain the possibilities. The drugs could be organized according to condition being treated. Both the condition and the drug would have to be entered, in effect enforcing the redundancy recommendation given above. However, once the condition was entered, the set of medications that might be specified could be automatically constrained depending on the condition being treated. Alternatively, drug choices could be constrained by the specialty of the practice, or the name of the physician, or even the history of the physician's pattern of prescriptions. With this kind of standardization, the doctor would select from a restricted number of choices. The automated system would use the information already entered to winnow the possibilities to choose from, reducing the possibility of error.

This kind of automated standardization would mean, for example, that a doctor prescribing pain medication for an arthritic patient would be unable to prescribe Cerebyx by accident instead of Celebrex if he has already specified that he's treating arthritis. Such a system, well implemented, could be a reasonable automation of the process we described earlier of redundantly identifying the drug with the condition. However, it is important not to constrain the independence of the physician too much. The system must have procedures by which the doctor can override the standardized set of options; otherwise, the doctor's limited choices might not allow exceptions necessary for specific cases.

One example where standardization does not appear to work is the conventional drug formulary system used by many health care organizations. Drug formularies are designed to limit the type of drugs that can be used.

This was supposed to save money by limiting the prescriptions to lower cost drugs, when there were roughly equivalent lower cost and higher cost options available on the market. However, studies suggest that such plans have had the opposite effect, increasing spending, while at the same time decreasing overall quality of care.[47] Among the reasons for this outcome are the need for doctors to go through special administrative procedures to receive approvals of exceptions, and the use of "second-best" treatments that later required further medical care.

The pharmacist's task and specialization

It is important to develop an understanding of task complexity to understand why solving the communication channel problem discussed earlier might not be sufficient for diminishing overall error levels. Let's take our best example of a method to fix the communication channel problem: writing both the indication and the drug. This approach seems like a very good solution for the communication channel problem and it might actually solve the prescription drug problems.

However there is a limit to this solution's overall effectiveness. Consider the pharmacist who receives the prescription. We mentioned in an earlier chapter that people have the capability of separating different types of information to different parts of their brain so that they can make composites. These composite states are the enabler of both creativity and, yes, of error. It is possible that a pharmacist would, therefore look at "Celebrex; Seizures" on a prescription but fail to notice the error. His brain may not see the incompatibility because of dissociation.

Right now the risk of this happening is not likely to be very high, but it's important to recognize that this could become a major problem if the complexity of drug prescriptions reaches a high enough level. The complexity might increase as the number of names of drugs increases to the point that there will be enough combinations of drugs and conditions to create confusion. It is also important to realize that the dissociation we spoke about varies from individual to individual quite a bit. So it is possible to select the people who are naturally (or by effective habit of action) good at making sure that both the drug name and the indication are consistent with the drug given. If a person makes an error, then we could reasonably consider whether improved training is needed or that someone else would be better at the job.

Still, what can be expected even from very proficient people is restricted by the complexity limit of the individual. Once the necessary tasks surpass

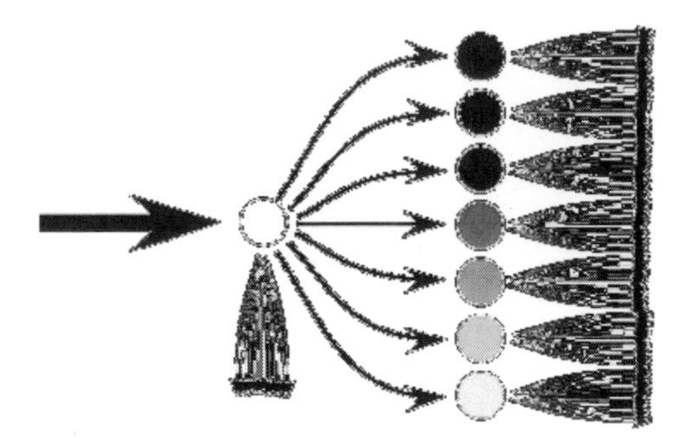

Figure 11.9: Medical routing system: The primary care provider receives all patients, makes some decisions and routes the others to specialists, who will be responsible for making further decisions.

this limit, we need a different solution, one which assigns the tasks to multiple people rather than to a single individual. This is what happens in specialization, which can take many forms.

The first approach is to divert cases into separate channels. Using this approach you can limit how many kinds of cases a particular individual deals with, reducing the complexity of his task. Specialization is a very important and effective technique for complexity reduction. We'll understand its importance more clearly if we take a look at the usual medical routing system.

Figure 11.9 is a diagram of a standard medical routing system. This arrangement isn't universal (the emergency room, for example, doesn't work like this), but it's still fairly typical. The white circle represents the primary care provider and the thick black line represents the many patients who come to see him. These patients have a very wide range of conditions, and rather than treat them all himself he refers them to specialists, the shaded circles.

The primary care provider thus deals initially with an extremely large variety of possible conditions. His task, however, is limited to addressing directly a more limited set of conditions and routing (assigning) the rest of the cases to the specialists. The specialists don't have to deal with the same level of complexity as the primary care provider. Each specialist receives patients with a much smaller assortment of similar or related conditions. The specialists allow the primary doctor to forgo treating certain patients, so that the actual treatment of the patient happens at a much less complex level. This makes a lot of sense—you're separating the cases so that the set of cases that any one person has to address is less complex. Still the

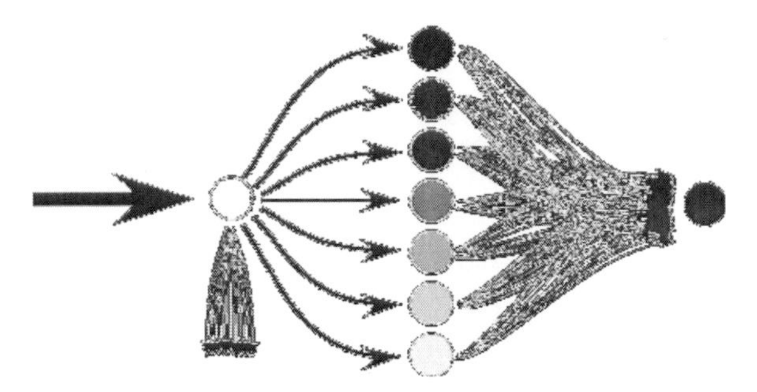

Figure 11.10: All specialists send patients to a pharmacist for medications (the primary care provider also does so; this is not shown in the figure).

overall process has a much higher complexity, which is clearly necessary to address the individual case needs. This is the point of specialization.

However, this diagram does not completely represent the entire routing system. Where do these cases go after the specialist? They go (via the prescription) to the pharmacist, the rightmost circle in Figure 11.10. This circle might also represent the nurse administering the medications to the patients themselves or providing other aspects of the care.

What's wrong here? The cases have been separated because together they're too complex for one doctor to handle. Now, however, they're reunited again (either in the pharmacy or in the care of the administering nurse). It's certainly not the case that all these patients now have similar needs—they still have vastly different treatment programs ahead of them. Of course, the full complexity of all of the cases does not fall on the pharmacist. There are many aspects of treatment aside from medication, and every condition does not require a distinct medication (consider the many different kinds of infection that are treated by the same antibiotic). Still, the routing system does reveal where problems are likely to arise. The architects of this system have applied complexity reduction to one part of the process, by referring patients to specialists, but have failed to do so at the other end of the channel, making it pretty obvious that the system's weakest point will be at the pharmacist/nurse, through whom treatments by multiple physician specialists must pass.

What's happened here is a good example of how systems adapt to increasing complexity. In general, as we learn more about how to treat medical conditions, the complexity of medical care increases because we learn how to effectively address a more highly varied set of cases using more specialized treatments. The community of physicians has addressed this increase in complexity by increasing the level of specialization, but the other parts of the system (like the pharmacy) have not found a natural way to adapt, so it's to be expected that problems will primarily arise at those points.

In order to make this system function more smoothly it is necessary to apply specialization to more than just one part of the process. From this diagram, the most obvious functions that should be specialized are the pharmacy and nursing care. There is already a limited degree of pharmacist specialization and nurse specialization. Nurses are specialized for emergency rooms, intensive care, anesthesia, and other areas. However, nursing specialization has been reduced in recent years with the cutting of costs, even as nurses' tasks have become more and more complex. The question is, how much specialization is appropriate? While it is clear that physicians have the most need for specialization, some degree of specialization of other professions in the tasks they perform is also likely to be necessary.

The importance of specialization can be found at many levels of organization. Specialization at higher levels of organization such as the care team or hospital would reduce the need for further specialization among the professionals who are working within that system. For example, we can consider the possibility of institution specialization, as found in children's hospitals, oncology (cancer) hospitals, trauma and burn centers. A pharmacy at an oncology hospital will be highly specialized for the very complex problems of drugs for cancer patients, for example. The existence of these specialized hospitals implies the importance of specialized knowledge reflecting the high complexity of care for patients in the categories (children) or with the conditions (cancer, trauma and burn) they treat. It also reflects the existence of a sufficient number of such patients to require a free standing institution. Developing specialized hospitals for every medical condition is not justifiable because the same circumstances do not apply, and because many patients experience multiple conditions.

At a lower level than hospital specialization, it is worthwhile to consider the strategy of forming teams. A team, consisting of doctor, nurse, and pharmacist (or a limited number of all three) can deal with the entire process of deciding what medication to use, filling the prescription, and administering the medication to the patient. If the unit is specialized to deal with certain types of cases, the number of distinct cases and possibilities that each individual has to deal with is drastically reduced. Moreover, different physicians even within the same specialty have different patterns in how they treat patients. This means that reducing the number of physicians that a particular pharmacist or nurse has to interact with reduces the complexity that they have to address. According to a colleague of mine, in Japan pharmacists tend to work with only a few local physicians. Such an approach (with the same set of possible medications) would lead naturally to errors being far less likely, since the possibilities for each pharmacist are drastically reduced.

Creating such specialty teams is not always practical. Still there are other ways to simplify the task of a pharmacist. The basic idea is to separate, as much as possible, the tasks into well-defined and distinct subsets, increasing the effective distance between the tasks even if they have to be performed by the same individual. One way to do this is to separate the pharmacy itself into different areas corresponding to physician specialties. If the specialty of the prescribing physician were marked on the prescription, the pharmacist would go to the part of the pharmacy with medications for that specialty.

The idea of using teams is also relevant when the standard form of specialization is not sufficient to deal with very high complexity tasks. Physician teams with collective decision-making and action are able to address much more complex tasks than individual physician specialists. Separating a single task to a number of specialists allows them together to perform tasks that can have the sum of the number of possibilities that each one of them faces. Setting up a physician team allows them together to perform tasks that have as many possibilities as the product of the number of possibilities that each one can address. This is a tremendously greater complexity. This is an ideal. It assumes that they all work in a mutually complementary way. Even without perfect complementarity, with proper training, they can work on tasks that are substantially more complex than individuals working separately.

How much specialization and collective action is appropriate—and for what specialties? There's no one answer to this question. Indeed, every hospital or clinic faces a unique flow of patients. The problem of specialization is also linked to the number of cases of a particular type that a medical system sees. Common cases should be treated in a streamlined way, at the other extreme, very rare cases should be treated as exceptions. The effort on a per case basis should increase gradually with how rare the type of case is. The formation of teams, therefore, combines considerations of efficiency and complexity. Specialization should be established so as to best fit the complexity of the medical care required.

What kind of success can we aim for?

Obviously a hospital cannot implement radical structural and organizational changes at once, and the impact of changes on costs is crucial. How to gradually transform an organization into the most effective structure for the complexity of its tasks is the ultimate point of this book and will be discussed in detail in Chapter 15. The problem associated with prescription errors, however, may be more directly addressed as it is likely to be first and foremost a communication channel problem, due to the convergence of multiple channels from different physicians to the pharmacist. As such, there are very small and easily implemented changes that can be made to reduce this kind of error. What kind of success can we hope for with these changes?

In 2000, the Institute of Medicine's report urged a national goal of reducing medical errors by 50 percent over the next five years. Many doctors and health care officials, even those who thought the IOM's medical error statistics were overestimated, thought that this target was overly ambitious. Indeed, today the reduction of medical errors seems far away. How does it sound to you? Well, let's roughly estimate what kind of reduction in communication channel errors we'd get, just using the simple technique of adding redundancy to recording practices.

Studies indicate that a patient admitted to a hospital has a 5–10% chance of being the victim of some kind of life threatening medical error. One simple change can reduce this quite drastically. If there are on average about 10 procedures carried out on each patient during this stay in the hospital, then there is roughly a 1% rate of error for a particular act. (This assumes that the errors are independent. [Aside: The way to calculate this is to ask what is the probability of no error occurring, which is $0.90 = (1 - .01)^{10}$]) Say we introduce some redundancy into the system by having the doctor produce two copies of the prescription, which are checked against each other before administering the drug to the patient. We assume that the errors in each copy are independent, so that each of them has the same individual error rate. By adding this one act (double checking the prescription), the error rate will be squared, and you'll end up with only a 0.01% chance of error for an individual procedure—and therefore a 0.1% chance of error for a particular patient. So with this one small procedural change, we have reduced a patient's chances of being subject to an error by 99%!

This is a simple calculation that does not take into account a variety of factors. Some of these factors would reduce the eventual error rate still further, others would increase it. For example, if the average number of procedures performed on each patient in the hospital is higher than 10, the reduction in error rates would be even greater. If errors are not independent because the people are too tired to write or read effectively then the error rate will be higher. Still the message should be clear: it is possible for a very simple change that addresses the actual problem to have a major impact on error rates, even making the error rate so small as to be unnoticeable. The conclusion is extremely important: the amount of redundancy that you have to introduce into a system in order to reduce errors to the point of undetectability is not large. We don't have to implement a whole slew of radical changes in procedure in order to dramatically reduce error rates.

Government agencies and independent health safety organizations have proposed lists of recommendations for changes to address medical errors and some hospitals have responded by spending lots of money and manpower on implementing many of them in a "coordinated attack" on medical errors. Other hospitals have become overwhelmed by the problem of implementing these recommendations. A special emphasis has been placed on technology and automation. It is important to realize that different recommendations will be appropriate for different hospitals, though some changes are likely to be useful for most hospitals. Differences in patient population, physician expertise and nursing programs, may result in a different space of possibilities for the same task at different hospitals—and the methods of error reduction that will be most effective will vary accordingly.

Though the choice of error reduction methods might be bewildering, the effects of appropriate ones, by this calculation, are exceptionally simple. All an individual hospital has to do, then, is pick one or two or maybe even three methods that address the particular communication channel problems they are facing. These changes should bring about a rapid adjustment and near undetectability in a short period of time. The moral of the story is that individual hospitals can try to implement reasonable changes and expect that they will lead to substantial and observable results. At the level of the individual hospital, 50% over five years is absolutely too modest a goal for reducing medical errors.

In the wake of the 2000 IOM report, the reaction of the health care and regulatory community was to focus its efforts on effecting change, originally through centralized action. This task was daunting, and ultimately unlikely to be successful, precisely because its goal is to produce recommendations and procedural changes that would bring all hospitals into line through stricter standards: standardized treatment policies and protocols, and technological devices that would reduce reliance on handwriting and memory. It's not that reducing errors by 50% over five years is not possible, but rather that because the medical system is a highly complex system effective change should arise from local actions, guided by an understanding of what goals can be reached and what approaches should be tried. Externally imposed standards and regulations will not result in a versatile system that can deal with the complexity of medical needs.

This places the onus on individual hospitals to test new ideas and evaluate them quickly. This fact has increasingly been realized. In March 2001, the IOM released a new report[48] arguing that the health care system had to be reinvented, through a "sweeping redesign" of the entire system—not via the imposition of a "blueprint" for care delivery systems, but through the creative implementation of new simple principles of care. As part of their effort to foster multiple promising routes for innovation, the IOM has now refrained from specifying proper procedures. It remains to be seen how the promise of the new decentralized approach plays out. Decentralization in and of itself does not imply effectiveness, but it is a step towards understanding how changes in the complex health care system can be implemented. Encouraging local experiments will allow innovative new approaches in health care to be discovered.

Conclusion

Although in this chapter, we've focused on errors concerning drug prescription and delivery, the basic insights are important for other kinds of errors in treatment, misuse or failure of equipment, and incorrect diagnoses: medical errors have to do with complexity. To dramatically reduce the incidence of errors, one must identify where the complexity arises and create a system that has adequately complex capabilities.

Using the notions of complexity and scale, we can gain a sense of what a successful medical organization would look like. In a successful organization, the convergence of messages from different types of individuals to one person is limited. Each communication channel is sufficient for the information flow. Unnecessary steps have been eliminated/automated where possible. Standardization reduces the complexity of tasks when it doesn't limit effectiveness. When complexity is unavoidably present, redundancies exist in the system to catch errors. The distribution of complexity across multiple individuals makes it possible for complex tasks to be performed effectively. More specifically, high complexity care is provided by teams with specialization of members of the team, as well as specialization of teams. More self-contained teams provide more individualized medical care from intake through diagnosis, treatment, release and follow up. In this way the traditional reduction of complexity by specialization at the level of the diagnosing specialist physician is maintained throughout the rest of the patient's care involving nurses, technicians, and pharmacists.

In this chapter we have focused on how we can think about and design medical services and the teams that provide them. It is quite hard, however, to understand the full complexity of these systems. Rather than designing them, the main role of management and policy makers should be to create an environment in which the systems create themselves. The traditional way to do this is through economic competition. For the health care system a different approach is needed, and this will be described in Chapter 15 on Evolutionary Engineering. The traditional approach to engineering and management by analyzing and specifying the system does not work. Both managers and engineers must use a new strategy based upon evolution to create highly complex systems.

PRELUDE:

EDUCATION

In 1997, a session at the International Conference on Complex Systems was organized by Michael Jacobson, who has performed research on whether complex systems ideas are understood by students and professionals. One of the speakers in this session was Jim Kaput, educator and developer of SimCalc, a system for teaching mathematics to middle school students. Together Jim and I approached the National Science Foundation about sponsoring a short conference and papers on the role of complex systems in Education. There were three areas that were addressed by this program. The first is the recognition that children could benefit from learning about complex systems ideas. The second is the idea that insights from complex systems may shed light on the process of teaching and learning. The third is the implications of complex systems research for understanding the education system and the efforts to improve education through system reform.

In the process of this meeting and discussion, I had the pleasure of working with others who devote their full attention to improvement of education and the education system. There are many who recognize the importance of complex systems ideas and have since then approached NECSI regarding potential involvement in projects that could be pursued to improve education. The discussions that we have held, led me to write these chapters on education and education system reform. I hope that they will become part of the dialogue on the education system along with the many other efforts that directly address issues ranging from individual learning to the effectiveness of school systems.

Chapter 12

Education I:

Complexity of Learning

Education and the education system

What is the objective of the education system? In the U.S., the purpose of the government is protection of individual rights. However, the public education system is often viewed as serving the state by educating proper citizens. For our purposes, we will encompass both types of goals: preparation of children for their roles in an effective society and serving children by developing their capabilities and opportunities for fulfillment. These goals can be viewed as reasonably consistent; after all, fulfillment of an individual is tied to many aspects of the appreciation and reward that society gives the individual, including but not limited to income. It is possible, of course to imagine systems in which individual fulfillment is counter to serving society well. We will argue that at least in a system that we will suggest, they are largely compatible.

In an earlier chapter, we came to the conclusion that organizations are increasingly too complex for hierarchical control to work. This means that society is undergoing or has recently undergone a major transition. At some point, not too long ago, organizations were less complex than the individuals who formed them. Today, organizations are often more complex than individuals. This transition is as dramatic, or perhaps even more dramatic, than the transition made to an industrial society approximately a hundred and fifty years ago. It is reasonable to suggest that the education of children will also have to change dramatically in order to serve individuals and society. The transition from the agricultural to the industrial age was accompanied by the transition from the one room schoolhouse to the schools we know of today. Such a transition has not yet taken place in response to the recent changes in society. Indeed, the education system today is an industrial era construct, modeled after a production line, manufacturing graduates.[49]

Production lines are designed to produce many copies of the same product. Still, we know that the society of the industrial era had many diverse needs. Many different kinds of workers and professions were needed. There was not only a need for workers that could work in industrial jobs, but also a need for businessmen, managers, engineers, doctors, lawyers, scientists, etc. If the educational production line were to produce only one product, it would not have served the society. We will suggest that the way the existing education system has provided for all these possibilities is by providing varied educations, often of unequal quality. While for an industrial production line uneven quality is not a good idea, the same is not true about the education system. The education system has provided a highly random experience with great variation due to the different conditions in schools and individual differences in teachers and students. This variation implies that children receive an education that is not necessarily a good one, but one that allows for variation in the outcomes, such that eventually children may be effective at different kinds of jobs. Today, people have become very concerned about the quality of education. Unfortunately, the current efforts to improve this quality are directed at increasing the uniformity and by doing so to increase the quality. This, however, is directly opposed to the needs of a diverse society. If the industrial era required a diverse workforce, the needs of today are even more diverse. We will discuss a different approach to improving education that promotes both high quality and high diversity.

Before we discuss the education system as a whole, however, we will consider education from the point of view of individual children and teachers. Our focus will be on the implications of the increasing complexity of society. If, as society becomes more complex, classrooms become increasingly complex, then students and teachers will experience a complexity that no individual can face successfully.

Complexity in the classroom

In general, one of the key points that we stressed in previous chapters is that the complexity of an individual must match the complexity of his or her environment if the individual is to be successful. Since a person is part of society, the environment of the person is formed by the society, just as the environment of a cell in our bodies is formed by the body. The environment of each cell (the fluid that it experiences in the body with homeostatic regulation) is not the same as the environment of the body (the air and the objects outside the body). Neither is the environment of a single person the same as the environment of the society as a whole. As society becomes complex, the environment of each individual cannot become as complex, otherwise the person would be overwhelmed. The environment of an individual must be simpler than either the society or the environment of the society for the benefit of both the individual and society.

The classroom is an artificial environment created by people. Like other environments we create, a key question that we should ask is whether the complexity of this environment is well matched to the people who are there, namely the children and teachers? An environment that is either too simple or too complex is not a good idea. If the environment is too simple, the people who are there will be bored in the short term and will decline in capability over the long term—a mind-numbing sensory deprivation effect. If, on the other hand, the environment is too complex, the people who are there will become confused or disturbed over the short term and will fail in various ways to meet challenges over the long term. An overly complex environment is an over-stimulating and ultimately defeating environment, very much like trying to play an unreasonably difficult video game.

This is an insight that has been observed, for example, in studies of home environments for the elderly performed by Alice Davidson, a professor of nursing.[50] Although order is important, as it allows people to make sense of their environments without experiencing disorienting confusion, it turns out that the environment for aging individuals should not be overly simple. A moderate amount of challenge and stimulation will cause an elderly individual to exercise mental and physical capabilities, which stimulates the maintenance and development of the neuromuscular system, maintaining activity and mental alertness.

In Davidson's study the complexity of an elderly individual's home was quantified by estimating the number of visually distinct environments (states of the system) that were possible given the objects in the home—clocks, windows, pictures, books, etc—and the various possible positions each object might take. Elderly persons living in environments of greater complexity displayed greater cognitive function and more robust circadian rhythms of locomotor activity. The study therefore concluded that elderly individuals' living spaces should provide sufficient complexity of stimuli. Complex, challenging and changing environments that are within the elder's ability, but require effort, stimulate the aging brain and body to maintain neurons, muscle fibers, and other tissues.

Similarly, information in the classroom environment challenges children. The transition to a highly complex society, implies there is an increasing danger from overly complex environments. The danger comes not only from the classroom itself but also from the entire environment of children who experience high information flow and high-demand activities including TV, Internet, and extracurricular programs, as well as the classroom activities. The concern here is not about whether these media are good or bad in direct influence, but rather about how much information flow they contain: the number of possibilities. Ironically, in some cases the high potential complexity of the environment may lead teachers to overreact and create overly simple environments. A tendency to oversimplify would lead to environments that do not adequately challenge or engage students. The often told story of the child who, through boredom, becomes a problem child is directly relevant here.

As we develop our discussion of the education system, we will be concerned with the differences between individual students. Ultimately we should recognize that for a complex society, individual differences are valuable far beyond their importance in the industrial age. This suggests that education must also promote individual capabilities and enable specialization. Fortunately, this is a positive development as far as individual students are concerned because of the increased care for the well being of each individual and for individual fulfillment through the development of unique skills and abilities. For the discussion in this chapter of complexity in the classroom, we should recognize that individual differences also imply that the complexity experienced varies between individual students. This means that we can expect that in a single classroom some will experience an environment that is overly complex, while others will find the same environment too simple.

Increasing complexity

The trend in the education system over time has been to increase demands on students. It is not that the demands are augmented as a single child grows, though this is definitely the case. It's much more important to note that demands on children at a particular age have changed over the years. The amount of material that students are learning at a given stage in their education and the number of different ways they are expected to learn is increasing.

There are a few factors that compound the complexity of the environment in the classroom today. These include many well-intentioned efforts to improve the system. Each of these efforts may be a good idea in isolation. However, they often serve to increase the complexity of education by increasing the number of possibilities that children may encounter. In order to benefit from these innovations, we must recognize the limitation of the complexity of children so that they are not exposed to an environment in which they cannot function effectively. In the following sections we will discuss three trends in innovation that naturally lead to increases in classroom complexity: finding innovative ways to teach advanced materials to younger children; integrating curricula; and bringing in new approaches that recognize individual learning differences.

Introducing material at a younger age

The first trend that increases complexity is the overall tendency to introduce new material into the education of younger children, usually by finding a new way to present material that was taught at a higher level. We are constantly discovering ways to bring college material into the high school or even the elementary school. Years ago, for example, set theory was a highly abstract topic that was only taught at the college level. Today it is taught at the beginning of elementary school. This was one of the components of a radical change in the math curriculum moving away from exclusively focusing on addition and subtraction to include sets, symmetries, patterns, logic, abstract thinking and word problems. We often find that with a little cleverness there are ways to teach material to younger and younger children.

There is also an ongoing tendency to increase the length of textbooks. There are various reasons why people want to add to the education of children. People who work on education often believe, nobly enough, that the most important contribution is to get children to learn more. People who have specialized knowledge often believe that this knowledge is valuable and that it should therefore be taught to children. Publishers want to sell new books and adding new material is an important aspect of an effective sales pitch. New material is a good reason for publishers to print books. How many books would sell if they said, 'we left out material from our last edition!' If we recognize a limitation in the ability of children to learn material, the real question is not what to add, but rather what to leave out.

Integrated learning

The second trend that increases the complexity of the classroom has been the introduction of integrated learning. This approach has features that are very consistent with complex systems ideas. In a more traditional education, subjects are completely separated from each other, for example, English, math, social studies and science would all be taught separately. Separating these subjects is like any form of decomposition: it ignores the connections between them and the many ways they work together in the world around us. Learning the ways they are connected to each other seems like a reasonable complex systems approach. One strategy is to have some integration by pairs of subjects. Another strategy is to have more general integration by learning subjects that have aspects of all of them. Still, we have learned that there is an advantage to subdivision and separation when we considered patterns and subdivision in the brain. Ideally, we need to understand the trade-offs between separation and integration in order to achieve the correct balance. The importance of this balance is not generally recognized. When we require each child to learn all aspects of an integrated curriculum, we are increasing the complexity dramatically.

While the topic of subdivision in the brain shows the importance of a balance of separation and integration, perhaps an analogy will provide a simpler explanation. Consider the way a house is organized into rooms that are often specialized: bedroom, bathroom, kitchen, living room, dining room. While the house as a whole should be integrated though the way these rooms relate to each other, still there are key advantages of the separation of the rooms. These advantages have to do with both the way the rooms are designed to perform their function, and the possibility that multiple people may be using different rooms for different purposes and would interfere with each other. Sometimes, people choose to combine certain rooms, like the living room and dining room and this can be appropriate for particular lifestyles and personal taste. More generally, however, despite the importance of integration of the rooms into a single home, over integration by making them all part of the same room leads to a loss of effectiveness. The balance of the two is important. In engineering today there is also a tendency to promote integration of previously separate components. Such integration often leads to tremendous increases in the complexity of engineering projects and many such projects have failed. This will be discussed in Chapter 15.

In teaching, some of the commonly paired subjects are math and English, English and social studies, and social studies and science. Consider the pairing of math and English first. One of the major innovations in the mathematics curriculum has been to teach it through English. Some of the impetus for this change is likely to have come from the SAT standardized test, where word problems are a major component. In order for students to do well on this test teachers taught word problems more intensively. Over years, the teaching of word problems has propagated down until it is often a central part of first grade math. The basic objective is to combine the ability to think in words and in math.

Why should this be a concern? Separating topics like English and math is a strategy of simplification related to the idea of subdivisions: consider separately things that are partially independent at first and connect them afterward. This means that children could exhaust the capacity of one part of their brains, the one more devoted to English, in an English class, and still have room for what they would learn in math class later on. Each subject would mostly affect a different part of the brain. This is related to the reason why children learn many different subjects in one day, rather than devoting an entire day to English, and all of the next day to math. They would go into overload if they studied one subject all day, which would manifest itself as a loss of attention and inability to focus on the material being taught.

Subdivision in the brain is an underlying strategy for effective functioning. It allows us to combine the parts together in various ways, without having to teach each combination. Likewise, the traditional separation of math and English is actually a very clever way to allow students to learn reasonably separate ways of thinking without causing overload. Once they are learned separately, it is possible to combine them together (e.g. by the connections between subdivisions of the brain).

This separation approach is not infallible, but the hidden trade-offs associated with combining subjects early on are not adequately recognized. Linking subjects together at an early age causes these ways of thinking to become more tightly bound (integrated!). This might seem like a good idea, but it doesn't take advantage of the natural role of subdivision and the possibility of various combinations. This can result in a loss of ability to learn, as opposed to an educational gain. Certainly, if a school attempts to teach all of the possible combinations of subjects, it will lead to a higher level of complexity in learning. There are many ways that subjects can be combined and integrated, which means that there are many possibilities and correspondingly a high level of complexity. When used well, integration provides an important aspect of the education of children. It enables children to see and learn the connections and various ways separated subjects can be used together. It also relates these subjects to the world around them, demonstrating the reality that the world is not neatly separated into subjects. The key point here is that integration is a good idea, but integration at an early age may expose children to too much of the complexity of the world.

One of the primary motivations behind teaching math through English is the idea that math is hard and by teaching it through English, we will allow more students to learn math. From the preceding discussion, however, we see that this approach may actually prevent students from learning effectively. Another motivation for advocating integrated learning is that many believe that word problems reflect the way the world works. According to this perspective, we always start to solve a problem by creating a word description that must then be translated into a math problem to solve. However, a better view of the situation is that math and English are different languages. We can start from either, creating a math description of the world can be independent of creating the English description. They are different ways of thinking about the world.

Of course, individual differences affect which mode of learning is most effective for any given person. Moreover, research in educational methods is always problematic because you can't evaluate children, teaching and schools independently of what you are teaching (a standard double blind experiment is not meaningful). Individual differences make it difficult to answer many of the questions that are central to evaluation of methods, such as: If you taught the same children in a different way how would they perform? People teach according to their beliefs regarding the principles of education, and there are numerous debates regarding what approaches work better. Oftentimes, the "winner," at least in the context of textbook publishers, is based more on politics than scientific evidence.

There is a different approach to integrated studies that allows for the limitation on individual complexity, namely integrated projects. Within a project each child could work on a different aspect of the problem using a different method. Such a differentiated approach to subject learning does not guarantee that each child learns the same as every other child; nor does it guarantee that each child learns all methods of learning. However, it does allow for differentiation and individual specialization, as well as team cooperative learning. Working on integrated projects can be a very good idea, in that it generates excitement among students and reflects the complex world in which we live. This approach to integration is quite different from the standard form of integrating subjects, in which all children are expected to learn all aspects and methods of study.

Focus on individual learning styles

A third innovation, the attention to learning styles, can be understood as a key effort to recognize individual differences and to increase the individualization of the education system. One way of dividing up the way people learn is through the concept of learning "modalities:" visual, auditory, and kinesthetic.

While children can generally learn in a variety of ways, and may learn in different ways at different times, it has been suggested that each child tends to learn best by using a particular modality. One might argue that the traditional school system tends to focus on a visual modality, with a secondary auditory aspect, and weak to non-existent kinesthetic aspect. Thus, children who learn kinesthetically are not treated well by the system. The existing system would consider visual learners as successes, kinesthetic learners as failures, and auditory learners would lie somewhere in between the two. It follows that if there are these different ways of learning, we should teach different children in different ways. However, rather than teaching individuals differently, teachers are using programs that attempt to teach all children in all ways. Carried to an extreme, this would result in children having to learn all of the material three times. Further, suppose we accept that each child is most effective at one of the modalities and less so at the other two. Then, if we evaluate children based upon their learning under all modalities, instead of finding that all children have become successful, we'll discover that all children fail much of the time. It's certainly reasonable to argue that all children should learn in all ways some of the time. Still, regardless of implementation, once we recognize different modalities and teach through them, we have multiplied tremendously the set of possibilities for teaching and learning and this is an increase in complexity.

The distinction between visual, auditory, and kinesthetic modalities is one example of individual differences in learning. Other discussions of individual differences have described the difference between analytical, creative, and practical capabilities[51] or the distinction between logical-mathematical, linguistic, musical, spatial, bodily kinesthetic, interpersonal, intrapersonal, and naturalist abilities.[52] There are thus various ways of learning the same information (as well as different kinds of information) and some children learn better using one approach as opposed to the others. The central problem is recognizing that for any one child—and for any one teacher—increasing the number of ways of learning increases the overall complexity of learning.

Symptoms of mismatched complexity

Together, innovations in bringing more materials from previously higher levels of learning, mixing modes and integrating subjects of learning so that they are combined rather than separate, and including more distinct ways of learning, increases tremendously the complexity of the learning that children have to do. It is important to emphasize that each of these is a good idea in and of itself and should be part of a good education process and education system. However, if they are implemented in a way that requires all children to learn in all ways, the total increase in complexity must be considered. As stated earlier and throughout the book, there is a limit to the complexity that an individual child can handle. If we exceed this threshold, we will do more harm than good. To solve the complexity problem, decisions have to be made about what children need to learn. This decision is central to the problem of education and education reform. Children can be taught many different things, but this does not mean that they should be taught all of them!

If there is a limit to how much individual children can learn in a given amount of time and we have reached this limit, what would we expect to find? We would expect to find that children are suffering from overload. This overload would not happen uniformly for all children. Even the children who are affected the most would not necessarily display the same symptoms. Still, there would be some common characteristics. If the amount of information provided exceeded an individual's capacity to absorb the information, the individual would resist the information by overlooking it, ignoring it, or responding ineffectively to it. We should look for problems that have been increasing dramatically over the same period of time as overall complexity has been increasing—the last couple of decades. In addition, we should look for problems that do not seem to be related to any specific circumstance or conventionally defined external cause because complexity exists in the diversity of what is going on, not in any particular thing. It is unlikely to be a coincidence that in recent years there have been increasing concerns about attentional problems, i.e., attention deficit disorder (ADD) and attention deficit hyperactivity disorder (ADHD). An attention problem is a problem with maintaining focus over extended periods of time—an impaired ability to concentrate on what is going on. An "attentional disorder" sounds quite reasonable as a model for what would happen if the environment were too complex, specifically if the amount of information being received were too high for an individual to deal with. Attentional issues do not appear only in children; adults have various ways they avoid excessive overload by not paying attention, from selective listening to a more global blocking of attention. If this were to become a general feature of how a child relates to the world, we would be right in being concerned. The discussion here suggests that we should be evaluating the complexity of the child's environment to understand this condition as opposed to just examining the child.

This informal discussion suggests that if complexity in the classroom is an important factor of education, and if children may be adversely affected by it, then the solution is to simplify that environment. It wouldn't necessarily matter what was removed, but it would be important to remove some things. There may be other reasons to choose one or other thing to remove, but from the point of view of complexity the choice that is made wouldn't matter. [53]

There are two main ways of decreasing the overall complexity; both dealing with restricting the types of environment experienced by children. The first method chooses a particular smaller set of things that all children will learn. The other method restricts the amount of information that will be supplied to each child, but determines what information to provide on an individual basis. This allows the children to specialize. Specialization in this case is similar to other organizational issues that we have discussed. The problem is to have an organization that is complex, while having each individual operate in a simple enough environment, so he or she is able to function effectively. Indeed, since there are different choices for what to teach, different teachers may reasonably choose differently. This is an important step toward thinking about the role of differentiation in the school system. Differentiation can serve a complex society that needs diverse and specialized individuals because the entire society is more complex than any one individual.

The amount of information that children have to learn is related to the complexity for the teacher in teaching it. The classroom may also be too complex for the teacher. The specialization of teaching different subjects (typically math, English, science and social studies) that occurs in many school systems as early as grades 5 and 6, and progressively increases from there, is evidence that teachers may have reached their individual complexity limit at roughly that level. Specialization, or the need for it, is always a clear indicator of complexity. The complexity of the individual teacher, as for any human being, has a limit, which makes them not complex enough to teach their students everything beyond a certain point. We can see this also in difficulties with training teachers to novel educational curricula or ways of teaching. It is most apparent in conversations with teachers who express the complexity of their efforts and in the common expression of "burning out" and leaving for less demanding jobs in terms of complexity.

It is quite clear that as society increases what it knows, the variety of possible skills increases, and there is a tendency toward increasing complexity of the environment of children in and outside of the classroom. The opposite danger, of overly simple environments, however, continues to exist. In some cases, as teachers adjust themselves to the increasing complexity of their environment, they may very well overreact and create environments for children that are overly simple.

Simple environments lead to boredom. Boredom is how we describe the emotional response. Over time, there is a cognitive impact leading to a loss of capability, simplifying and "dumbing down" the responses of the individual. Ironically, from the point of view of the education system modeled after the production line, these consequences are positive as they make subsequent education easier by making children more uniform in capabilities. From the point of view of the individual or the current highly diverse and complex society, whose effectiveness is based on the existence of highly diverse and capable individuals, this is a negative development. Other consequences are also, however, possible. For example, boredom leads individuals to "act out." Acting out is a natural and direct response to boredom: an attempt to increase the stimulation and complexity of one's environment.

Up to this point in the chapter, we have discussed the classroom and the environment as a context that is out of the control of the individual child. In general, people have some control over their environments and tend to exercise this control to make their environments better suited for them. In this regard, it is important to realize that the classroom and the educational system, in general, is a coercive environment. This means that the environment is imposed upon the children as opposed to allowing the children to develop or choose it. Essentially, we judge students based upon how well they fit into this coercive environment. A broader perspective suggests that when a child finds that she is in an environment that is not suited to her, her natural response is to try to change that environment. In the case of a child that is overwhelmed by an overly complex environment, this kind of response is less likely because of the challenge she faces in dealing with the environment, but it could occur. In the case of the child that finds herself in an environment that is too simple, the natural response is apparent: an attempt to make the environment more interesting, more stimulating—so-called disruptive behavior. Therefore, disruptive classroom behavior is actually quite a reasonable consequence of a coercive environment. If an adult were required to spend every day in a boring environment, and there was no choice about whether to be there, it would be increasingly likely that the adult would attempt to disrupt that environment. It is not accidental that we often identify the "bright" kids as those that are disruptive. Of course, efforts to escape that environment, to be somewhere else, are also quite reasonable responses to such a situation. The nature of the coerciveness of the education system can be directly measured by the system's response to such behaviors.

It is essential to realize that individual differences imply that in the same classroom one child may find the complexity overwhelming, and another may find it boring. These capabilities are likely to vary by subject or by differences in teaching/learning style as well. How can a teacher respond to this circumstance? Teachers often target their classroom activities based upon their philosophical approach. Frequently the approach is to "include everyone"—a least common denominator approach, causing those who are able to learn more to be bored and disruptive. In other cases, the approach is to "challenge the class" targeting the best and elevating those who are not too far below in capability in that particular subject or that particular method, but leaving others tuned-out and overwhelmed. Rarely are teachers able to build an environment that can accommodate the many different abilities and ways that children learn in their classroom. This is easy to understand, the complexity for them (for the teacher) would be too high. Teachers can learn how to design classroom activities and learning programs that are better adapted to the existence of individual differences in the classroom. Still, it seems clear that the ultimate solution to this problem must be found elsewhere, perhaps through changes to the overall structure of the education system.

The solution of this problem is tied into the other issues we are facing with the education system. Therefore, we need to turn our discussion back to the education system level and consider the concepts of education reform, so that we can finally address the question of what should be done for students and teachers. Indeed, looking at the system from local and global levels of organization is consistent with the general multi-level and multiscale approach of complex systems.

Chapter 13

Education II:

The Education System

Crisis in education

The crisis in the education system is becoming, or perhaps already is, as acute as the crisis in the health care system. We can analyze the education system problems with the same tools we used to consider the health care system. There are similarities and differences between the two, but the main conclusion is basically the same. Today, we are using an approach that would be well suited to a large-scale highly redundant system. However, this approach is inappropriate as it is being applied to a complex system. The resulting consequences will likely be disastrous. Has this approach worked so well for the health care system that we should do it again with the education system? It has taken 20 years for people to realize that the health care system may be going in the wrong direction and with the education system there is a much longer period of time before feedback is received. Children have to grow up and become part of society before we can really evaluate whether we have done a good job of raising them. Therefore, it is unlikely that we will be able to learn from our mistakes for a very long time. This is where the ability to learn from experience in the military or health care systems, guided by the principles of complex systems can be particularly helpful.

In this chapter we begin our discussion of the education system itself, by considering how the system is organized from a complex systems perspective. Then we discuss problems with the system. While there is common agreement that problems exist, we find that the problems have been misidentified, and this is leading to an incorrect approach to improving the system. Both the evaluation of the problem and the solution are built upon a large-scale uniform approach to what children need to learn. Consequently, the solution being pursued now would lead to a much more homogenous population. By contrast, the problem that we will identify is different. Thc approach we recommend to improving the education system based upon complex systems insights would lead to a much greater variety of individual capabilities. This would be better for both the society that needs to fill highly diverse professions to promote innovation, and for the individuals that would be able to advance their individual talents and interests.

Education system organization

The first thing to notice is that the education system is remarkably distributed and has very weak interdependencies. From my experience in visiting schools, it is safe to say that in many schools what is going on in one classroom has very little to do with what is going on in another classroom, even the classroom which is right next door. If a dramatic event happened in one classroom, such as the replacement of one teacher by another, there would be little if any change in what was going on in the next classroom. Moreover, classrooms in a school can appear quite different from each other. This is even more the case when you consider what is going on in different schools. For example, teachers in one school do not usually know what is happening in another school in the same district.

The education in different classrooms, however, often shares common textbooks, guidelines for what children should know, and aspects of teacher behavior. It is important to distinguish this indirectly imposed uniformity from interdependence. Whether or not interdependence exists lies in the answer to the following question: When something changes in one place, does it affect the other? In the school system it generally does not. Therefore, commonality imposed by external forces, such as commonalities associated with the education of teachers, is not a sign of strong interdependence.

As we have discussed many times, interdependence is useful when system function requires a dependency. Otherwise inessential dependencies can get in the way of effective function. It's true that when we think about complex systems we think of interdependencies, so in some sense you might think this independence suggests that we shouldn't think of the education system as a complex system. Still, the lack of dependency is significant from a complex systems perspective precisely because people tend to consider the education system as a system rather than thinking about individual classrooms as the essential unit of organization. When thinking about the education system as a whole, therefore, we should not forget that the classrooms are largely independent from each other.

From the point of view of the complexity profile, the aspects of classrooms that are the same give a large scale uniform behavior, while the independence of the classrooms implies there are substantial aspects of the system behavior that are fine scale and local. This is not accidental. Why would this be the case? The education system is organized this way because key aspects of its task are fine scale and local. Specifically, the highest complexity task in the education system is the educational relationship between the individual teacher and each individual student. Many people conceive of education as a process by which a teacher gives a set of information to the students. If this were the case, then teachers could give the same information to many students. It is known, however, that the quality of education decreases dramatically with classroom size, especially once the classroom size is more than a few over 20. This is one indication of the importance of the individual relationship between teacher and student. There are many other indications that can be understood by considering the dynamics of classroom activities. The interaction between the teacher and the students in a classroom is highly complex, involving the unique qualities of the teacher and the unique qualities of each student.

The complexity of the school system's task also comes from its duty to prepare students for the complex world in which we live. The variety of different professions requires many different skills. With the complexity of society increasing, the complexity of education must also increase. If we accept for the moment that the task of education is a complex one, we can understand why strong dependencies among activities in different classrooms are not a good idea, and thus why the system is designed in this way.

Still, there is a danger associated with having classrooms and schools largely independent of each other, and that danger is—tremendous variation in the quality of education provided. Random quality is the natural outcome of the lack of dependency between the different classrooms. This does not seem like a very effective system. Before we address this issue directly, let's consider what those who are concerned about the education system say about the problems and develop our own understanding further.

Identifying the problem

It is widely understood that it is time to improve the U.S. education system. A helpful next step is to think carefully about the problems that exist, and to properly identify their origins within the system. It would also be good to recognize the successes of the current education system, so that we don't eliminate the strengths along with the weaknesses.

Many of the concerns about the education system arise in the context of concerns about the U.S. economy's effectiveness in global competition with other countries. In 1983, the National Commission on Excellence in Education issued a report entitled "A Nation at Risk," which was a call to arms to improve the education system in the face of international competition. They stated: "Our Nation is at risk. Our once unchallenged preeminence in commerce, industry, science, and technological innovation is being overtaken by competitors throughout the world."[54]

Despite these dire warnings, the U.S. economy today is considered the major driver of the global economy. The Gross Domestic Product is about $11 trillion, more than 1/5 of the global economy, which is estimated at $51 trillion. The population of the U.S. is only 4.6% of the population of the world, less than one in twenty. On a per capita basis the U.S. economy is larger than all countries except Luxembourg.[55]

Still, according to the results of standardized math and science scores (specifically, the Third International Mathematics and Science Study (TIMSS)), the U.S. is far behind many countries in the world today.[56] The results, if anything, have become worse over time rather than better, for children graduating from high school. The U.S. average score is significantly below the combined average of the scores from all other countries taking the test.

If we look at these scores as indicating success in the future, this seems to be a paradox: We have had many years of low scores, and still the economy is remarkably strong. How should we interpret this paradoxical situation? Does this mean that math and science have nothing to do with economic success, or that it has an inverse relationship to success? Surely this seems unreasonable. Should we conclude that we do not have to worry about our education system? This is also not justified. However, claims that the education system is in trouble because of the poor performance on these tests are also unjustified considering the evidence just presented.

Rather than any of these explanations, I would like to make a claim that is based upon a fundamental measure of a system that is important to complex systems: the variety of its members. A system performs well in facing complex challenges when it has high variety. We can understand this in the case of the modern economy and technological and corporate innovation. While a standard science and mathematics education might be good for some purposes, creative use of that knowledge is likely to arise when people take many different approaches to its use. The economic growth that arises from innovation occurs when people are exploring many different ways to do things. These different approaches naturally arise when people learn in many different ways rather than only one way.

Moreover, it is also quite clear that math and science are not the only skills that are important in the modern economy. For example, success in building an effective company involves many skills that are not taught in math and science classes. We can start from the observation that computer programming is not in the classic school curriculum, neither are the basics of engineering or management. Are the best software programmers the ones who do best on standard math and science tests? This is far from clear. Are the best web page designers the ones that do best on these tests? This is even less obvious. What about corporate executives, managers, actors, musicians, sports stars, how do they perform? Lest there be questions about the economic importance of some of these categories, we might remember that movies, music and sports are a major export of the US and make an important contribution to the economic activity in this country.[57]

Indeed, it seems that very few of the highest paid professionals hone their professional skills by taking continuing education courses in mathematics. We can push this issue even further. If science and math were key to effectiveness and success, then a popular book regarding success in the world, such as Covey's book *Seven Habits of Highly Effective People*,[58] would surely contain a significant fraction of scientific and mathematics problems for people to cover as preparation for dealing with the world around them, but it doesn't. Of course, the book you are reading now is about scientific ideas applied to the real world, but this is clearly not the usual case, and the nature of the science discussed here is not the same as that which is taught in conventional high school mathematics and science curricula. Conventional biology, chemistry and physics do not seem to warrant ongoing studies among the general public, unless they are specifically targeted for individual professions. For example, physicians may study particular biological subjects.

As the author of an advanced textbook with many equations, I would be very pleased to have more students who can understand advanced math. Do I think that this is what everybody needs to know, however? Absolutely not! This book is about ideas that are widely applicable in the real world, do I think that everybody should know them? No. There are many different things that people can and do do in the world, and there are many things that they need to know to do them. There is a greater amount of important information in the world than any one person can know and it is good that we don't share all the same knowledge because then we wouldn't be able to contribute differently to making our world work. Is there something that everybody should know? Perhaps, but making this the only role of schools will surely not be a good idea.

If our school system is desperately in need of improvement, it is far from clear that by creating a generation of students who perform well on TIMSS, we will succeed making the system a success. Today we have a society that requires outstanding educators, scientists, engineers, managers, physicians, nurses, technicians, psychologists, writers, computer programmers, lawyers, soldiers, accountants, designers, artists, musicians, actors, athletes and many others. Striving toward a single ideal, such as success on standardized tests, may actually reduce the overall effectiveness of our society.

The importance of variety for the effectiveness of society is not the whole story. We can also consider what is good for the children themselves. What would make an education system that serves each child well? Would it be one that fits him or her into a mold, a single ideal, or one that provides attention and opportunity for his or her interests and capitalizes on his or her strengths? Children are quite different from each other, with different skills, personalities and desires. The ideal of opportunity does not mean that children should be the same. When we ask essential questions, such as "What about the fulfillment of a particular individual child?" we come to recognize the importance of both the individual and the society and the interplay between them.

How should we understand the variety that arises in the context of the existing school system? We can think about variety as arising from the variation among schools, among individual teachers and among individual students. The interplay of these forms of variety gives rise to the variety of educational outcomes. The issue of variety is also entangled with uneven quality.

Where can we recognize the constructive role of variety in the success of the existing U.S. school system? One of the keys to this puzzle is something that many people say about themselves: their career was set on course by a single teacher. This suggests that in a highly variable system, almost all teachers are not going to be right for any one child, but it is enough for one teacher to be the right one for there to be success. There is then a central problem of matching the right teacher to the right student. If every child were to have the same right teacher, the solution would be easy, we would have to identify particularly good teachers and only use them. However, this is not the case; different teachers are good for different students. This does not mean that there aren't teachers who are good for many more students than others, but recognizing the variability is important. In the existing system, having different teachers with different teaching styles and little standardization might just give enough chance for students to end up having the right teacher at least once, so that the overall system will be a success even though most learning is only so-so.

The variety that arises due to differences between schools is often related to socioeconomic conditions and the characteristics of their community context. There are great disparities in the education system. Typically, suburban schools are considered much better than those located in the inner cities. Notice, however, that we are now talking about variations in quality, not variations in the specifics of the education provided to individual children. Variations in quality are not necessarily what we had in mind, but these are still a form of variation. They reflect the reality that the U.S. education system is highly variable in many ways, including quality.

The variation among schools, and particularly the low quality of some schools, leads to an important motivation for school reform that is distinct from the issue of international test score comparisons. There are manifest problems with many of our schools. They often provide a poor environment for learning: buildings in poor repair, overcrowded and insufficient facilities for teaching, widespread alcohol and drug use, violence, crime, etc. Of course, this is not just a characteristic of the school, but rather a property of the environment in which they are located. The classic problems with inner city schools, and more generally with schools in poor communities, are easy to notice and have been hard to solve. Indeed, some people argue that the key to good schools is socioeconomic condition, and that targeting those conditions will be more important than targeting the schools. Better schools, however, can offer a means by which people can improve their socioeconomic conditions one generation later. Therefore, even if context matters to education, we can still justify focusing on improvement of education as a way of improving the situation over all.

From the society's perspective, it might be possible to argue (whether cynically or realistically) that having poor schools and poor schooling for some students may have been good for the industrial era society, in which a significant fraction of low-skilled workers was necessary. (What would have happened if everyone were trained to be professionals?) Today, however, as we change to a post-industrial society, and despite the current profusion of low-skill service jobs, this seems like a bad idea both for the children and for society. From an individual perspective, it is clear that children are not achieving as much as they can.

The variation in quality, both between schools and for any child between classrooms suggests that we should be concerned about and improve the quality of the education system. What is unfortunate is that the recognition of these problems of school educational quality suggests to many that uniform education is the solution. Indeed, the high profile that is given to the comparison of TIMSS test averages with those of other countries obscures the extreme variability that is the source of much of the constructive educational opportunity.[59]

Moreover, the education system quality depends on how much people care about it locally. When the baby boom generation was in school in the 1950s and 60s, people cared a lot about the schools. Once they left, in many places the number of students decreased until it rose again when the baby boomers' children reached school age. This led to several decades in which education was not the focus of societal efforts. Few new teachers were hired. Salaries increased in other parts of society but not for teachers. Therefore, it is not surprising that the education system needs attention at this time.

The reason it needs attention, however, is not poor results in standardized tests, but rather that random quality is not good enough. The problem of the education system is that the quality is highly non-uniform. What we really want is a system with high quality everywhere, while still retaining a high variety. Having only one good teacher over 12 years of learning seems to be too low a success rate. This is particularly true with the increasing demands of our complex society. Before we present a complex systems approach to creating a system that will have high quality and high variety we will describe the current approach to education system reform.

Current educational reform: standardized testing

Today it is generally recognized that schools need to be improved. It is possible, as we argued above, that some people have at least in part misidentified the problem. What is more important, however, is understanding that the accepted approach to the solution, a large-scale uniform approach, is inadequate and inappropriate for addressing the problem.

The dominant approach to education reform relies upon an indirect way of evaluating the schools and compelling improvement. Rather than directly evaluating a particular school, all students are given high-stakes standardized tests. The idea of this approach is that because of the dire consequences to students of failure, teachers and administrators will be motivated to improve their educational programs so that children pass these tests. It is not too hard to make an analogy between the uniform action of tank divisions and the uniformity of the standardized testing approach. If this strategy is fully implemented, it will not be long before all students around the country are taking the same test at the same time. This seems to be a clear case of a large scale and uniform approach to education.

It is reasonable to expect that a strong force like this will have an impact on the school system. However, given our understanding of complexity, it will ultimately only serve to change the way the school system is failing our children and society. As one might expect with a uniform approach to a complex problem, initially there are likely to be misleading successes. This is just like the initial successes of the U.S. in Vietnam or the Soviet Union in Afghanistan (and the initial success of the U.S. in Iraq as compared to the current problems). Initially there was a great deal of success: territory was won and troops occupied the countries. In time, however, the failure of the approach was apparent. Increasing numbers of troops were killed in a large number of small military actions taking place over years, many civilian lives were lost, and the local society could not function effectively. Eventually, the entire military force withdrew in great frustration. This is how large scale approaches to complex problems fail. At first, the large scale approach seems to be working and its impact may be felt, but over time it fails in the details, piece by piece. Over time, these pieces add up to form a disastrous failure. Therefore, in education, we can expect that there will be many signs of success, which will give advocates of the standardized testing approach an opportunity to declare victory and reinforce their convictions that they are doing what is right.

Standardized testing is not new. Originally introduced in 1901, for years the SAT test has served as the primary hurdle for entry into colleges. Still, there were many ways around this test that provided alternative options to students. These included attending vocational schools or not going to college. Some states, e.g., New York, have had a state-wide standardized test for high school graduation. Today, however, many more states are introducing tests as high school graduation requirements and there is a new federal mandate for annual standardized testing.

Standardized tests promote "teaching to the test:" narrowing the scope of education to include only the material that will be on the test and teaching how to take tests. This makes sense to the people involved, since their success will only be measured in the taking of tests. In order to evaluate whether this is a good idea we also have to ask: Is the taking of tests really an effective measure of capability in students? Does it indicate eventual success or fulfillment in society? By looking around at how our society works, it seems quite clear that test taking itself is not what people do and not what they are paid for either. This is not a good sign for the usefulness of tests as measures of success. From the point of view of individual adaptation, if test taking is the only way we measure success for students and for schools, then we should expect that this is what students and schools will become good at over time.

There are other ways the system adapts to such a measure of success. Schools that are evaluated by the average score of students find ways to prevent students who will score poorly from taking the tests, or even "cheat" by giving answers to the students.[60] Limiting which students can come to a school or testing only some of the students at the school is a time-honored way of improving the average score of students in the school. How different is teaching the specific material that will be on the test from just giving the answers? If what we really wanted our children to know is only what is asked on the test, at the time of taking the test, then this would be enough.

The motivation behind standardized testing rests on the assumption that schools will teach children generally useful knowledge and that the test will evaluate their learning in general. This assumption would be true only if the following conditions were met:

- There exists a standard body of knowledge that is useful.
- Testing provides an unbiased measure of this knowledge both through its content (what is asked) and through its process (test taking itself).

Specifically, the evaluation of taking the test must really be a direct measure of the knowledge that is necessary.

Even if these conditions are met, it is still not clear that having children do well on average on the test would make the education system successful in its objectives. Success depends on what the objectives are. For example, even if children can learn useful knowledge and that knowledge is tested objectively by the test, it doesn't mean that having all children learn the same information will make the society effective. If having an effective society is the objective of the education system, the system would fail.

Many of the problems associated with standardized testing are well known. Why, then, do people think that they are such a good idea? Some may believe that tests are good ways to evaluate students. However, others may use them because they do not recognize that there are other options for improving the system. In the next section we will discuss a number of such options. Before we do so, however, we should consider the possibility that the motivations for imposing standardized testing and increasing its stakes may actually be different from what people claim motivates them.

One of the possible motivations for the movement behind using standardized tests is to oppose the development of distinct subcultures. Prior to the movement for standardization, there was a movement toward developing educational programs in other languages (particularly Spanish) and providing more cultural flexibility in education. While the U.S. has a strong tolerance for individual differences and for cultural diversity, there is a limit to both. There is a strong belief in the need for a common language and common frame of reference for interactions and shared ethical context. This does not mean that people think all aspects of culture should be the same, but it is generally held that there should be a common base that enables the functioning of the social system. Identifying this common base is difficult. Still, from the societal point of view, one of the key functions of the educational system is to provide this common base. What should this cultural base consist of? There is a lot of ambivalence and many strong opinions in this country regarding the answer to this question.

Standardized testing does not address the complicated subject of culture directly. Instead it promotes the idea that there is a certain crucial set of information and skills that children should know. Because this knowledge does not appear to be "cultural," such claims are more politically palatable. Still, if the requirements include most of what is possible to teach children in the allotted time, then educational uniformity precludes cultural and other kinds of diversity. It seems reasonable that many people feel a common cultural base is important. The right balance between cultural diversity and common culture is unclear. The point that we are making here is not to evaluate the proper balance. Instead, the discussion in this chapter suggests that serious problems will arise from standardization.

The current approach to improving the education system based upon standardized testing is headed in the wrong direction. Rather than having many good teachers, it will ensure that children will have a more uniform education without that key element of success: the really good and unique opportunity. This strategy runs counter to the underlying direction of increasing complexity in society and will ultimately fail to give children the opportunity to flourish through their individual capabilities.

Once we have recognized that the education system could be much better, but that the existing approach is deeply flawed, what can we do to improve it? There are many people who care about the education system and have worked hard to develop new ideas for what and how to teach children. From a practical point of view, the independence of individual classrooms helps us understand why many strategies for improving the education system have been unsuccessful in their efforts to "reform" the entire system. Since the key to the effectiveness of the school system resides in individual learning, intervention efforts have to work locally. This means that people will have to devote much of their effort to local success in a particular classroom. People expect that if success happens in one place, the results will inspire others to adopt their approach. However, if classrooms are largely independent this does not happen easily. This creates a great deal of frustration among people who are working in this area and who expect to impact the entire school system.

While standardized testing remains the dominant approach today to education reform, it is not the only one. People are working on other approaches, which range from charter schools to home schooling. Some of these approaches are supported by a complex systems perspective. Our next step is to consider what the complex systems perspectives suggest. Our understanding of high complexity in education suggests that there isn't one right solution, there are many for different circumstances—the right one is not the same in each place or for each child. The key is in recognizing that achieving high quality does not require sacrificing variety. By using ideas from the study of complex systems, one can arrive at solutions that will increase both quality and variety.

Identifying solutions

Complex systems insights provide a number of directions for improvement of the education system. There are several aspects of this solution that bear emphasis: the need to intervene locally; niche selection instead of standardized testing; and the selection of teachers, schools, and learning environments. In general, these approaches are motivated by recognizing the complexity of education of individual students as a local task and by recognizing the complexity of society as manifest by its dependence on diverse and specialized knowledge rather than a common knowledge base that is shared by all individuals in society.

The need to intervene locally rather than globally

The mantra of activism is "think globally, act locally," and it applies well to lessons from the study of complex systems. The significant independence of individual classrooms and the complexity of the task immediately indicate that each place has a unique situation. Recognizing what is necessary to improve a classroom requires recognizing its own circumstances. Here is where the solution from the military actions in Afghanistan shows the way: Special Forces. There are many problems in schools across the country. We should recognize that different problems in different places require local and specific actions.

The solution is to develop teams, "Special Forces," that can identify specific actions that should be taken in different places. There is no one solution to all of the problems. If there were, the large force approach would work. Instead, there may be need in different places for: smaller classrooms, after school help for children, financial supports, firing of teachers or administrators, big brothers or sisters, information technology, innovative teaching methods, repairing schools or building new ones, special interest programs, or teaching parents along with children, just to name a few. As with military actions in complex environments, large scale standardized testing will fail where Special Forces will succeed.

The possibility of multiple criteria for success

Evaluation is an important part of any program that is goal-oriented. If we care about education we do have to evaluate how well we are doing. By rejecting standardized testing, I do not argue that we must reject all forms of evaluation and assessment. While it is not necessarily true that evaluation of the education system is the same as evaluation of the progress of children, we do need to grapple with both of these if we are going to make progress. The evaluation of children (and evaluation in general) seems to bring almost everyone to the notion of standardized tests. After all how can we evaluate if we don't have standards? Some suggest that the alternative is individualized assessment: evaluating the progress of an individual against their own goals, i.e. against their own potential. This often sounds good but has its own problems. How can we measure against a child's own potential when we don't know what that potential is yet? We have clearly not reached the point of being able to evaluate in advance what a child is going to be able to do (otherwise we wouldn't have any problem with evaluation or education!) and unfortunately complex systems indicates that such prediction is not going to be easy.

Where can we look for other ideas? Evolution provides an important analogy: "niche selection." As we discussed earlier, the competition in evolution takes more than one form. There are many different ways to be successful. Think of all the variety of plants and animals, different climates, ecosystems, and different roles within each ecosystem. Each of them is subject to competition. Being good at one role is quite different than being good at another. This would be like having many different "standardized" tests, though the concept of test here is too limited because there are many forms of possible evaluation or competition. Each of the competitions could be very rigorous, but there would be many different ones. This is actually what is going on in society today. There are many different professions that a person could choose. Still, in order to do well in any of these professions you have to excel. Whether you are a baseball player, a basketball player, or a golfer, you have to be outstanding at that sport. Even within a sport you often have to excel at one of the possible positions (in basketball: center, point guard, coach), not all of them. If you consider a businessman, doctor, lawyer or scientist and look within each of these professions, you find many different "niches" in which people excel. Of course it is much harder to establish multiple measures of success than one measure. This, however, is compatible with what society is doing and the education system is preparing children for this society. If we consider what has happened since the agricultural age and through the industrial age, we see there is an increased need for specialization.

Whenever the topic of specialization in education comes up, a question that often arises is "isn't there some need for everybody to know certain things?" Perhaps there are some things that should be learned by all children. However, the discussion of standardized testing today is about improving the schools, it is not about what should be known by everyone. It would be quite reasonable for there to be a discussion of what should be taught to all students, and this might include aspects of common culture discussed above. However, this discussion should not be confounded with the efforts to improve the education system.

It might be argued that standardization makes sense early in education and the key issue is the timing of specialization. Historically, specialization begins in high school and becomes more specific in college. Informally, it has been apparent for years that many successful adults learn unique skills early in life. This is particularly true about professions where there is high degree of competition for a few cherished spots, for example, in professional sports or in the performing arts. If we had much longer lives and could afford a first education that would be common to all and afterwards more forms of specialized education, then maybe more standardization would work well. Even then though, the argument is doubtful. Early education is important to developing the key connections and processes of the brain. Moreover, when we think about what people in different professions do, we see the wide diversity of skills and capabilities they have. What makes one person successful in one profession is remarkably different from what makes a person successful in another. Education should provide an opportunity for children to flourish at what they are good at. This should involve early rather than late specialization of teaching in order to accommodate the differences that exist between children in how they learn effectively, what they learn readily, and what they require greater effort or time to learn. At the simplest level of analysis, in a uniform system in which all children learn at the same pace, and all are learning the same material, the rate of progress is dictated by the slowest rate of progress among all children. If children are allowed to learn at different rates, then children progress rapidly where they can, and more slowly where necessary, and the rate of progress for each child is greater than the uniform case. Since different children do better with different approaches to teaching the importance of individualization is even greater. Where diversity of outcomes is important, the usefulness of rapid progress in areas of individual talent is even more apparent.

Indeed, the standardization of education is a lowest common denominator process that suppresses the unique abilities of individuals in order to have all individuals reach the same place at the same time. It is a mass production approach to child development. To appreciate how limited this approach is, remember how different children are from each other. If you don't remember, spend some time with children. For any parent who has more than one child, this is quite apparent!

What would multiple niches look like in education? How many ways are there to evaluate effectiveness of a child that are relevant to professional success? Start from conventional standardized tests. There are several different subject areas. We could identify more than only one combination of scores as metrics of success. Doing better on one test might compensate for doing less well on another. A minimal standard on each test might still be used. However, this minimal standard would not be the only measure of success. There is no particular universal truth in considering a particular balance of English, math, history and science as the only way to be successful. (The opposite extreme is embodied in the spelling bee where just one area of English, spelling, has been identified as the unique arbiter of success.) Surely, we can come up with other ways of weighting different subjects to measure success. Adding a variety of other tests, with the testing format still being standardized would greatly increase the possibilities for a variety of evaluation metrics.

Broadening the perspective as to what constitutes an evaluation would be next. For example, the use of timed tests is an issue. Some criteria for professional success depend on whether one can act rapidly, others on whether eventually one does the best job regardless of how much time it takes. This suggests that in addition to timed tests, we consider evaluating the set of works that a child does in writing, projects in math, science or social studies, and works of art. This is often called portfolio assessment. Then, we might also be concerned about quite different aspects of what affects success: interpersonal skills, emotional maturity, empathy, effective setting of goals. These are all aspects of what has been called "emotional intelligence."[61] Many believe that such skills are more important to professional success than high standardized test scores. Believing that such abilities cannot be evaluated is unreasonable. If some of the evaluations are not as cut and dry ("objective") as tests, so be it. Just because standard tests are objective does not mean they are valid measures of success. Once we devote attention to these methods of evaluation, we have to consider how to put such evaluations together to set criteria for success. If developing wide ranging evaluations and criteria for success turns out to be a difficult task, it is reasonable that it would be. Identifying appropriate and diverse metrics for truly evaluating children for effectiveness in our complex world is a crucial task that we should devote substantial effort to.

The role of selection

There is a second key evaluation that we need to consider: the evaluation of the education provided to children. Many think that this is the same as evaluating the children—if the children do well then the school is doing well. However, this statement only makes sense if we have standards that can compare different children receiving different educations and assume they are similar enough such that we can compare their progress and use that as a metric to compare teachers, schools or education systems.

To evaluate the education system, we could also devise many standards. One approach is to evaluate progress against some, possibly diverse, measures that are used to evaluate the children. Another involves observations of the dynamics of progress of children as they learn (formative assessment). Both are relevant to evaluating the learning environment. Still, the main issue in improving the education is not the evaluation itself, but the possibility of selection. Selection is the powerful force for change in evolution. This is where competition (comparative evaluation) plays an important constructive role. Today, and for many years, the process of selection has played almost no role in the education system. This is where the education system is very different from the health care system. Indeed, it is the reason that at the local level there are quite different problems in the two systems. In the health care system, despite the medical errors, we are not usually critical of the doctors and of the doctor-patient relationship, while in the education system, we are often critical of the teachers, schools, and teacher-student relationships. The most natural solution to this problem, and really the most reasonable solution, is to allow for selection in the education system.

If you think about it, parents have more choice about the mechanic that fixes their car than the teacher that is with their children for most of the day and most of the year. If a mechanic fails to meet your needs, you pick another garage. There's no corresponding ability in the case of your child's education. This is an astonishing feature of the education system.

School choice has become a politicized topic in the context of public and private schooling. School choice as it is currently discussed concerns whether public money can be used to support private education that might have religious components. This would give parents a kind of choice, but the choice that I am suggesting here is simpler but yet more powerful, the possibility of making any out of a number of choices: teachers, schools, etc, even if all of the choices are within the public education system. Today schools and school systems are very protective of their ability to assign teachers. They are truly afraid of the actions of parents in making choices. After all, too many might choose one teacher and not enough choose another teacher, putting the entire system into turmoil. Well, isn't that the whole purpose of selection? Educators also often believe that parents do not know what is best for their children. Of course selection does not always result in the "right" choice. Evolution has a significantly random aspect to it—but this randomness is important for creating the possibility of change and progress. Selection is the mechanism through which progress occurs.

There are examples where selection is beginning to gain legitimacy in the school system. The existence and policies of charter schools and other special schools are beginning to provide the ability to select schools and teachers. An essential component of any school improvement program, therefore, would be to provide an increasing number of options. Preventing choice may be safer for the people who currently exercise control within the schools, but it inhibits the system from progressing in any substantial way.

Summary of the big picture: education and society

The existing education system is based on an industrial model of a succession of stages (grade 1, grade 2, …) and a product release at the end (graduation). The standardized tests that are the main tool of education reform today are part of an industrial-era approach to quality control. These tests measure the quality of the product at the end of the production line (or at check points along the way) according to standards so that they all come out the same. There are many ways to modify the education system so that it will be more compatible with a post-industrial age society. It may be that the entire education system will change more dramatically than any of the possibilities we discuss here. Still, even relatively minor tweaks, which are suggested or supported by complex systems ideas, would have a major impact on how education and the education system perform.

The key to the perspective I argue for here is to expand the role of the education system: the education system of the future must identify *which* education a child should receive as much as it should provide that education. The identification is a mutual selection process of the education system and the child /parents.

In order to provide different children with different educations we can't just create alternative programs. We also have to continually evaluate/assess children to determine how to direct them through the system. This is a complex routing task, with the objective of giving young people a smooth transition between education and work. (With continuing education it is not even clear that the transition happens as a real transition.) Evaluations, while they may be quite individualized, also have to compare children against other children. Niche selection implies that we have groups of students that are similar enough to allow for comparison. Children also should be grouped together to simplify the education task of teachers that cannot teach highly diverse and specialized educational programs. Still, whether we bring such children together in the same place at the same time for such evaluations or whether we compare them indirectly through electronic or other means, the existence of some comparative process is generally necessary to evaluate/assess both the children (for the purpose of routing them) and the education that is being provided to them.

We have considered two different directions, from the point of view of the complex environment of each child and from the point of view of the failure of the education system as a whole. Both perspectives lead to the same conclusion: the education of each individual child should be the focus of reform efforts, not a uniform strategy. The approach to individualization and selection of the proper environment for each child is a great challenge that will require major modifications to the education system in order to succeed. Still, this is an important aspect of the transition of the society away from the industrial age where mass production in the school system was an at least partially reasonable model, to a system that can prepare children both for individual fulfillment and for the complex society in which we live.

There will be trade-offs, not just positive effects of this transition. This discussion is not about idealism—it is about realism. One of the cherished ideals that may have to be abandoned is the idea that we all share a common way of thinking about the world and can talk and communicate with each other. This ideal does not seem sustainable. Individualization and specialization are ultimately not compatible with shared communication. The idea of a complex collective as a society does not mean that we are all friends of each other. It means that we function effectively together by being different in complementary ways. Local communication should be effective, but global communication may be limited. This doesn't mean we shouldn't try to keep some degree of commonality by maintaining common aspects of education. Still, this may be increasingly difficult to retain as society becomes progressively more diverse through specialization.

We also noted that the complexity of the education process led us in past to have highly independent classrooms, even though there was standardization from the textbooks and teacher training and expectations. If we switch to an approach of dealing with individual specialization, many of the similarities between educational systems in different places will be reduced. At the same time, the interdependence will increase through the need to route children to various possible educational programs and allow choices to be exercised. This implies that classrooms will become less independent of each other. Competition between children and evaluation of them by niche selection, as well as competition between classrooms for students will increase the interdependence, but this interdependence will be local—not global across the education system. When we consider the complexity profile, we see that the interdependence reduces the very fine scale complexity, increasing the complexity at a scale just above. Decreasing the uniformity of the education system reduces the complexity at the highest scales. All in all, the effect is to increase the complexity at intermediate scales consistent with a move away from the "either-or" conventional approach—either completely uniform or completely independent—to system organization. From a complex systems perspective, it is apparent that this change in the complexity profile is necessary for the education system to create a generation that will form the society because the complexity profile of the system has to match that of the society that has already, as well as will be, formed.

The education system has a responsibility to teach students what they need to know so that they can become successful citizens. Such success enables both fulfillment of the individual and flourishing of the society. Objectively, there is no fixed concern about the role of the individual or that of society, but rather the compatibility between the two. A successful education system will enable children to develop their potential and to fulfill their potential in their roles they come to play in society.

There is another topic that we have not discussed except in passing in this section, but that has particular importance. In society today, individuals are not the essential unit, but rather various groups, teams and organizations form the functional units. Thus, the education system must recognize that educating individuals in and of itself is an incomplete task. Even if we consider the role of individual education in enabling individuals to form effective teams later on, this is not enough. Ultimately, the teams themselves become the result of the educational process.

Traditionally, much of the development of the ability to work effectively with others is considered outside of the school curriculum. The skills of cooperation and social interaction are often learned through games and sports, where constructive interactions are necessary, whether in a competitive individual game or a team sport. Gaining the ability to cooperate continues to be an important part of team learning. However, as we have pointed out in the context of the health care system, developing professional capabilities of teams requires a much higher level of integrated functioning to perform highly complex tasks. These have to be gained in the context of specific professional interactions and in a manner that is effective for the particular individuals involved. Just as having individuals learn a sport like basketball is insufficient to enabling a particular team to play well together, learning the individual skills are not sufficient to enabling functioning teams in other contexts. While it is clear that early education is not yet the place where professional teams should be formed (similar to the too early integration of parts of the brain in integrated subjects), still attention should be paid to the learning of team skills and the progression through education that ultimately gives rise to teams must be recognized.

Prelude:

International Development

In 2000, Alberto Bazzan, the director of leadership development at the World Bank, asked me to prepare an educational program on complex systems for World Bank executives. The goal of the program was to provide executives with basic insights and concepts from complex systems that would be of general use in management, not just for the mission of the World Bank—providing assistance for the alleviation of poverty and economic development around the world. In the process of preparing this presentation, however, I inquired and learned about the current state of World Bank programs. What I learned suggested that there were some positive developments that were bringing World Bank activities toward what would be suggested by a complex systems perspective, but that there were other aspects of the strategy of the World Bank that would severely hamper its progress. This understanding forms the basis of the discussion presented in the following chapter.

Soon thereafter Alberto was embroiled in some internal World Bank politics (which itself is quite traditional) and became engaged in other management responsibilities. In recent years, the opportunities for NECSI to become involved in development efforts have increased through contacts made with other individuals working at several international development organizations, including both the World Bank and Asia Development Bank. Renewed interactions with the World Bank have come through the invitation of Frannie Leautier, Vice President of the World Bank and head of the World Bank Institute. Interactions with Asia Development Bank are growing through the initiative of economist Hans-Peter Brunner. My initial exposure to World Bank activities has served me well in establishing a basis for discussions about current development community activities and what should be done to reach development goals. It is increasingly apparent that global efforts are progressing from more traditional approaches to incorporate complex systems ideas. Still, a more complete understanding of complex systems ideas is necessary to increase the effectiveness of such applications.

I remember that even as a child I was concerned about starvation due to droughts and endemic poverty in Africa. The World Bank and other development organizations have been concerned about these problems for many more years and have devoted great efforts, financial and human, to addressing and resolving these problems. The limited progress achieved is sufficient to motivate the use of complex systems insights in this context.

When I, as a child, considered possible professions, it always seemed that the highest contribution would be a contribution, however small, toward solving these problems. I had, however, from an early age determined to be a physicist. My solution to this paradox was to contribute to knowledge that might ultimately bear on economic development and thus on the elimination of global poverty. I chose a quite practical area of physics for this reason—the study of defects in electronic materials. I was greatly disappointed when practitioners making silicon computer chips did not view my efforts as necessary, because they used pre-existing heuristic rules. Over time, I chose to study more fundamental areas of physics and ultimately the study of complex systems, which I viewed as a fundamental rather than applied area of inquiry. It was surprising to discover the relevance of my fundamental research on complex systems to practical problems in the real world. To find a relevance of this effort to Third World development is a fulfillment of a great dream!

Chapter 14

International Development

Introduction: Third World development and complex systems

Achieving the goals of development in much of the world does not just mean economic growth, although that is often a big part of it. Development is a complex term that is often an umbrella for many different processes—relieving extreme poverty and hunger, decreasing violence, improving access to education and health care, utilizing untapped resources, reducing the incidence of fatal diseases and decreasing economic disparity. Basically, the goal of development is to foster the growth of a functioning and productive society. As you might expect, in the complex systems view of development, the many aspects of this goal are connected to each other in complicated ways. However, only relatively recently have the main players in international development incorporated this key insight into their practices.

To take a complex systems view of development requires two insights that aren't typically fully understood. First of all, it is crucial to understand how each kind of intervention affects the internal structure of the country. Some interventions that seem obvious—like providing relief food to regions suffering from famine, as we'll discuss below—can have disastrous consequences if not done right, worsening the situation instead of improving it. Other kinds of intervention may have no effect at all because of the stability of the status quo. Second, complex systems research tells us that if the goal of development is to create a functioning society, then it simply *cannot* be planned from scratch. All the relationships and interactions that make up such a complex system cannot be known in advance, so development cannot simply involve drawing up a blueprint and then implementing it. If you can't plan what you're going to do, what can you do? A lot, it turns out. To explain, we will start by looking at approaches to the real and urgent problem of food aid. Then we'll look at how these issues have played out in the past activities of one of the world's largest supporters of development activities: the World Bank.

Food aid

The basic mission of food aid is to provide temporary relief of food shortages in poorer regions of the world. It keeps people alive under conditions that otherwise could lead to death. If done right, food aid can prevent the irreversible effects that result from even short periods of malnutrition. By itself this is readily understood to be a positive objective. The paradox of food aid, however, is that it sometimes leaves its recipients hungrier and less empowered than before.

In 2002, a United Nations sponsored assessment of food aid programs in Northern Ethiopia concluded that food aid in the mountain ranges of Gondar had worsened food insecurity and hindered human development.[62] With regular food assistance the population in the region was able to grow steadily, increasing the strain on local agriculture and the natural environment, while never relieving the need for external aid. This is a reasonably natural outcome of food aid that is provided for an extended time. Other examples, however, are clearly counter to this one. India whose food aid was once a major global concern (e.g., in 1960 and 1966) has become self-sufficient in grain production (though within the country variation is great), despite a growing population. People continue to debate the benefits of food aid and problems of population growth today.

From a complex systems perspective, there is another problem with many programs that provide food. The standard practice involves distributing food directly to individuals, who usually come to a distribution center to receive it. This form of direct food aid has a subtle, but ultimately disastrous effect: it disrupts local mechanisms of food production, gathering, and distribution. These local social and economic structures are bypassed entirely by the distribution of food relief, which weakens and sometimes destroys them—leaving the region even more vulnerable to food shortages and even more desperate for continued aid. This problem is often compounded by the migration of people from rural areas to urban areas, where they can be more easily reached by food aid. Such an exodus naturally disrupts the existing mechanisms of food production and distribution even more, making them even more difficult to reestablish.

The paradox of food aid also appears in the efforts of aid agencies to avoid corruption and exploitation of the programs. In general, those who deliver food aid are focused on making sure that everyone receives the food in an equitable fashion. They devote substantial resources to preventing people from diverting food supplies in order to sell it to others for personal gain. The paradox here is that this kind of commercial behavior is what characterizes developing economic activity and social coordination—the very basis for the development of an effective economy.

It is worth pointing out that developed countries are much less concerned about equitable distribution of food in their own countries than they are when they're distributing food aid. Indeed, many wealthier countries have extensive populations of poor and malnourished individuals, despite the wealth of others. What is the justification for this in the developed countries? It is hard to justify ethically. Still, this is how what we consider to be a functional economy operates. The extreme opposite situation is easy to understand. If everybody has the same resources, there is no reason for an economic system to exist. If we don't admit the possibility of inequality and impose a highly effective external distribution system on impoverished countries, we are doing exactly the opposite of what is necessary to create a sustainable system; we are destroying the socio-economic system that exists.

The problem at the heart of these paradoxes is that delivery of food to each individual separately creates a flow of supplies that does not require any internal social structure at all for its fulfillment. In societies with healthy food distribution systems, food passes through production, processing, packaging, handling, distribution, storage, and sale, with each step managed through relationships and transactions between different kinds of workers. Conversely, in a community receiving direct food aid, most of this process is carried out by outsiders, individuals not belonging to the community. Thus there is no environmental pressure requiring social organizations to provide food. Therefore, it is quite natural that no internal system of food distribution will develop—and any existing vestiges of that system will atrophy. This is one of the important reasons that so many famine-gripped countries find it difficult to escape famine conditions even after the disaster (natural or man-made) that initially caused the food shortage is long gone.

Imagine, on the other hand, that the supplies were delivered heterogeneously—to some more than others, or to a limited set of entrepreneurs in different amounts. This situation would require trading and redistribution through interactions within the system. Such interactions are, in some sense by definition, the structure of the economic system. By stimulating exactly the kind of interactions that food aid officials strenuously decry, this program would have taken steps towards creating a more complex economy. Still, just creating any socio-economic system is not a good idea. A much better way to do this would be in a way that is consistent with the preexisting distribution system of the society, one that is consistent with its natural environment, by inserting supplies as close to their natural origins as possible. This would be much more effective in retaining the existing socio-economic structure. Although the individuals who donate their resources to food aid will see individuals exploiting their donated resources for financial gain, we must realize that the effectiveness of any economy includes people finding resources by chance (claiming land or inheriting it, for example) as much as by hard work, and selling it to others for financial gain.

Many food assistance programs have had an immediate negative impact on the social organization of their recipient countries and a long-term negative impact on economic development. The problem is that the underlying social structures of the country have been weakened, so that consequently there's not much to work with when trying to strengthen economic structures.

Does this mean that we should not provide food aid at all? Of course not! However, food assistance agencies must realize that providing aid at the individual level impacts the societal or community level of organization. Providing direct and long-term aid to individuals may seem like a good deed, but it is at the expense of necessary social structures that are difficult to reestablish. This kind of food aid starves the forest by feeding the trees, so to speak.

It has long been an objective of forward thinking people to help countries develop means for supporting themselves even beyond their immediate need for food. This is the role of development agencies. To look at how development agencies have gone about the problem of strengthening the country as a whole, we turn now to discussing the activities of the World Bank.

The World Bank

In 1944, the World Bank, or the International Bank for Reconstruction and Development, was created to promote economic development in nations around the world by providing low-interest loans and grants to developing countries. The idea of its founders was that if the Bank invested well in successful projects and programs in these countries, the resulting financial growth would enable a return on investment through interest paid back to the bank on the original loan. The development model that underlay this strategy relied on large-scale infrastructure projects, like dams, to stimulate the economy of the recipient country.[63] In the case of large dam projects, the goal was to provide the country with a means of producing cheap hydroelectric power. The power could be purchased and put to use by businesses, stimulating local economies and creating jobs. This was supposed to result in a mutual-win situation: the country would develop and money would be paid back on the loan.

Historically, these loans have performed well enough, so that the World Bank has made a profit and can continue to exist. However, these investments have had very little impact on development and many parts of the developing world—including those in which the World Bank has heavily invested—remain in poverty. Large dams, in particular, have frequently failed to provide their claimed benefits and have at times exacted a cruel cost on local economies due to displaced populations and environmental problems.

The Comprehensive Development Framework (CDF)

Over time, recognition of these limitations led to the reevaluation of the programs of the World Bank, particularly in the late 1980s and into the 1990s. James D. Wolfensohn, President of the World Bank since 1995, wrote a letter in 1999 that significantly altered the course of the World Bank's activities.[64]

Wolfensohn's new approach, called the Comprehensive Development Framework (CDF), identified the problem of development not just in terms of economics, but rather in a more "holistic" framework. Viable economic development could only occur, he argued, when there is an improvement in key social institutions (government, judicial, financial institutions and social programs), human conditions (education and health) and physical infrastructure (water, sewers, energy, transportation, environmental protection). Underlying his statements was the conviction that economic development occurs when people have the means and capability to build the society.

An important second aspect of the CDF was to relinquish the notion of development as a unilateral activity on the part of the World Bank. The World Bank had been extensively criticized as dictating the same solution everywhere, with no consideration of the local concerns and issues. Instead, the CDF called for local governments, non-governmental organizations, religious and community organizations and private sector agencies and philanthropists, to join discussions with the World Bank regarding the necessary steps to take for development programs in each country. This consortium of "players" in the development business would work together to promote development, rather than depending on the central control of the World Bank for decision-making or coordination.

The key to implementation of this new "framework" was what Wolfensohn called the "matrix of development activity." The matrix was to serve as a management tool, a way to plan the development activities in each country by entering projects into an organized grid. The development "players" would collaborate to design a matrix for each country, making sure that all the institutional, human, and physical necessities for development that the CDF had identified were covered. With assigned roles and responsibilities, each agency could then go about implementing its own part of the blueprint.

The planning trap

From a complex systems perspective, the CDF's assertion that development was a holistic process was an extremely positive change of attitude, as was its recognition that there were many "players" in the development field. Despite its many important insights, however, the CDF did not go nearly far enough in incorporating complex systems insights. It has major flaws that have and will continue to undermine its progress if not addressed.

The CDF's first flaw is what I call "the planning trap."[65] World Bank interventions in the third world are often for countries whose socioeconomic system is functioning at a very low level. These countries have very weak infrastructures, social supports, and economic activity. For these countries, the idea of planning investments on a matrix in order to foster development amounts to planning the structure of a functioning society. This might be reasonable if we could understand exactly how a functioning society operates, but social systems are highly complex and it is not possible to plan such a system, much less plan the process by which we could go from a non-functioning to a functioning system.

A loose analogy might be the problem of putting together a person out of organs, a kind of Dr. Frankenstein task. Getting each part to work effectively and connecting them to function together when they depend on each other in intricate ways is harder than it looks. We may very well be aware of some features of functioning systems, and we could certainly identify these and work to put them in place. However, even if we can identify the parts of a complex system, this does not guarantee that we can put them together to form a functioning whole. Assigning one associate to an organ would still not enable a real world Dr. Frankenstein to succeed. This means that coordinating development activities cannot be as simple as adding entries to a grid.

System and environment

The other failure in the CDF's approach is much more subtle, but just as fatal. One of the most important insights of complex systems is that you must understand the relationship between a system and its environment. In development, this might mean the connection between a country and its wider global context, or its most generous benefactor. In other words, it's not enough to say that there are many interconnected parts of a socio-economic system that are important for effective functioning. If you only consider the connections within the system, you're missing half the picture. Just as important is the nature of the connections between the system and its external environment. Development will occur only when a country's environment promotes it *and* the internal mechanisms of the system can enable it. Individuals in the development community often recognize the importance of the interactions of the system within the environment. The subtlety here is that the CDF as a framework for action does not put this recognition at the top of the list of priorities in considering the impact of interventions. It is a question of emphasis, but ultimately a crucial one.

The natural environment includes geography, natural resources, and climate. Consider the simple observation that much of the undeveloped part of the world is equatorial, not just in Africa and Asia but also in the Americas. The problem that the equatorial region creates for development is not just the high temperatures. This part of the world often has drastic changes in climate that can last for years.[66] Extensive rains often alternate with severe droughts. The equatorial region is also where severe floods, typhoons and hurricanes are more frequent. All of these have an important impact on the possibility for successful development. Radical changes in climate make it much harder for infrastructure and social mechanisms to persist. When we consider that floods, hurricanes and fires can have devastating effects in the U.S. and in other highly developed countries in the temperate zones, it is reasonable that environmental effects might still today be a key factor limiting development in the tropics. Thus, severe weather may even be sufficient to disrupt development if it is not prepared for adequately.

The hottest topic in development studies today is the idea of "sustainable development:" will the development process we promote in a region right now be sustained into the future? The popular understanding of this concept usually focuses on ensuring that development does not disrupt the *natural* environment by destruction, pollution, consumption, over-exploitation or waste production in a way that creates problems for future generations (and for the environment itself). From the above comments, we can see that sustainability should also be concerned with whether development can be sustained across many different circumstances and events that can happen; specifically the effect of environmental changes that happen not infrequently in the equatorial region.

The natural environment is, however, not the only environment we need to be concerned about. Development must also leave the system with an ability to exist in its human environment—its own cultural setting, that of neighboring countries and the larger global system. The global economic system may provide opportunities for development through trade and commercial interactions of various kinds. However, the global economy is a rapidly changing and often turbulent system. It is difficult today for well-established large companies and even developed countries to survive effectively in this environment (for example, consider that one of the largest and most successful economies in the world, that of Japan, has been in great difficulty in the past decade). How should we expect development to be persistent and sustainable in this context? Development in most of the modern "industrialized" nations occurred in a very different global environment than the one that surrounds developing nations today. We cannot simply assume that the same structures and mechanisms that worked to industrialize 19th-century Britain, for example, will work to develop 21st-century Nicaragua. Development strategies must be consistent with the rapid changes that are taking place in the world.

The flip side of this problem lies in understanding how a system becomes dependent on connections to its environment. Much of development aid is intended to be part of a temporary intervention. The idea is that an external party, such as the World Bank, will provide assistance for a short time and then walk away, leaving a developed nation fully able to support itself without the continuing intervention of the external party. Development agencies conceive of financial assistance with a goal of fostering independence rather than dependence. Development agencies should therefore act as the scaffolding that helps workers build a house, which can then stand on its own when the scaffolding is removed. Although this is often the intention, it doesn't always happen. The problem is that a scaffolding around a fledgling complex system—one that adapts by its nature to its environment—will almost always end up entangled with the system as it develops. Almost any intervention will create a dependency between the system and the intervener. If a country's economy is dominated by generous loans from the World Bank (which must be the case if the loans are to have a large impact on the nation's development), that economy will become structured as a response to that intervention. It's thus very possible that development assistance will make the country being helped completely dependent on that aid, rather than helping it become an independently viable (sustainable) system on its own.

This immediately gives us a paradox: How can one help when help creates dependency? To address this paradox requires recognizing that different forms of help lead to different kinds of dependency. We can surely work to minimize the dependency by choosing the form of aid carefully. Sometimes, the benefits of help must be weighed against the problems of dependency. In other cases, the dependency may actually be a positive rather than negative effect. Indeed, it is often recognized that the objective of modern development efforts is actually to create a dependency between a local economic system and other parts of the global economy that are mutually beneficial. In this case, we can think about the aid as transforming to gradually become the desirable interactions between the system and the environment, thereby becoming an integral part of their relationships.

It is reasonable to suggest that development that is sustainable has to be "natural" in some sense—natural to the people who are part of the developing society and natural within their context. Natural means that the ultimate role and the path to that role are consistent with the abilities and strengths of that system and the context in which it is located. This is a reasonable condition to set if we want sustainability. Local differences will also make development different in different places. Often it is assumed that globalization is identical to uniformity, but this is not at all self-evident. When we pay attention to the differences between places and people around the world, this is clearly not the case. In general, successful development in one country will not necessarily look like success in another country.[67] A complex systems perspective suggests that the diversity of parts of a system is important for its ability to withstand challenges. Assuming that mankind will face challenges in the future, internal diversity will be extremely important. This should not be interpreted to mean that any conditions are acceptable in the interest of diversity. Still, it is possible to improve conditions and to create societies that function better in many diverse ways.

In an important sense, this returns us to a well-known paradigm of development. For various economic reasons, there are countries or areas (e.g., parts of India, Southeast Asia and China) that have been developing rapidly in recent years largely because of their own initiatives and the initiatives of the people or companies that are investing in them for economic gain. The interactions that result, flows of capital, products and people, are intrinsically compatible with the global economy because they are created by it. They may not always be stable, they are surely changing rapidly, consistent with the rapid changes in the global economy itself. Still, even if this process is not always consistent with the current notions of sustainability, they appear to reflect a robust development process.

There are objections to this form of economic development due to the possibility of exploiting the environment or the people. One of the major problems associated with development, for example, is companies treating the workers poorly: "exploiting cheap labor." Unfortunately, the steps of development that have occurred in the currently developed countries were also fraught with many environmental and human costs. We may hope and even demand that these problems will not be repeated in their full severity and that transitions between development stages will occur more rapidly. Still, given the existing experience, it may be hard to know how to bypass them completely.

This discussion might lead people to the conclusion that there is no aid that is good aid. Such a statement is too extreme. If there are places where development is taking place without special help, and there are places where development is not, surely there are places among the latter where a small amount of help ("a nudge") will enable development to take place. The key is to provide the help in a way that leads to the natural process of development, not to attempt to direct development in a fully planned out way according to an artificial set of constraints or objectives. The nudge approach would enable the system to organize around a permanent rather than a temporary solution. This is an important feature of a constructive realization of globalization, in which developing countries are able to develop because of long-term, mutually beneficial interactions with the rest of the world, rather than due to temporary aid. To the extent that temporary solutions are provided, the time for which they are provided should be severely bounded to prevent organization around these solutions.

Another tension between the system and the intervening agencies has to do with the process of change in a networked system. Both functional and non-functional societies are networked systems. The interdependencies in each system give rise to its stability in its present state. The system's status quo, no matter how dysfunctional, is consistent with the interactions that regularly take place within it and those interactions are consistent with the status quo. This means that any intervention that is designed to really alter the way the system works will be naturally opposed by interactions which restore it to where it was before.

The problem can be visualized as a simple physics problem: moving a ball out of a valley. If you give the ball only a slight nudge in one direction or another, it will simply roll back to its original state at the bottom. We say that this system is self-consistent. This stability is essential for an effectively functioning system—otherwise any small push or change of the environment would cause major disruptions. However, even systems that we want to improve have this self-consistency, which means that we have to apply a sufficiently large force in order to cause major change. When we apply this larger force, we are inherently destabilizing the system and the process of change will not be smooth. If we push the ball with enough force to get it out of the valley, it could go over a hump that will cause the system to "run away." The probability that it will end up in a valley where we wanted it to end up are small indeed. Instead, it will likely end up somewhere else, in a state that's less functional than we want—a complex system has many more possible states that are dysfunctional than well-functioning. This leads to the crucial point: in a forced development process, we should be prepared to apply at least as great a force for stabilization as we do for change.

How do such instabilities manifest themselves? They appear in the form of social violence that undermine development and can result in total disruption, or various dysfunctional rather than functional societies. The forms of violence may range from coercion and exploitation, all the way to civil war. The most direct stabilizing forces that are often required are the strengthening of a country's police or military, or the direct imposition of order through external police or military actions. Still, it is important to recognize that strengthening these stabilizing structures is likely to carry its own cost in terms of impacts of intervention. Strengthening them might easily lead to a dictatorial police state, a scenario that has also occurred frequently in the third world. Our discussion of providing stability is like saying a functioning physiology must have bones, but how much of the system should be bone? This question naturally leads us back to the underlying problem with planning. If a system does not develop on its own, we have to understand the development process enough to provide the right forces to make it work—a truly complex task. An essential step forward is to recognize the multiscale complexity of the society and thus of the task involved. This understanding goes hand in hand with recognizing the limits of the planning process and what to do about it.

Aid at different levels of organization

In a sense, the underlying problems with food aid and development aid as they have been traditionally performed are complementary. The first addresses the survival of individuals, while the second considers an entire country. Neither of them adequately recognizes the multiple levels of organization that make up a society, the many scales at which society functions.

Food aid is administered directly to individuals. This disrupts the higher levels of social organization. Development aid has historically been provided in the form of loans to (or under the control of) a few individuals at the level of the country as a whole. These individuals are then responsible for the deployment of these funds in a way that stimulates financial growth in the economy at large—often through large public works projects like dams and reservoirs. The benefits, then, are supposed to "trickle down" to lower economic levels. The problem is that these tactics have rarely been shown to address poverty directly. This approach is very much like our discussion of financial flows in the U.S. health care system in Chapter 10. Since the objective of the flow is to disperse it to the parts of the system to achieve individual well-being throughout the society, we have the problem of taking a larger flow and breaking it up into many small flows, resulting in turbulence. Turbulence can be seen in undeveloped parts of the world through various forms of social disruption, including violence. When the major resources that a country has are centrally directed, the likelihood of such disruption is high as individuals or groups attempt to gain control of those resources. This is the opposite extreme from the problem of food aid, where food aid is too often distributed at the individual level at the expense of socio-economic growth.

In order to understand what kinds of activities will relieve hunger while also retaining the existing social structures or planting the seeds of a self-sufficient society, we need to consider the relationship between individuals and their society.

Humanitarian aid agencies that respond to hunger must think of their task as not only relieving individual hunger but also fostering internal mechanisms in the targeted region so that it can begin to serve its own community. These two projects should operate on very different time scales. When a large social disruption occurs, such as a natural disaster or a civil war, there is an immediate need for aid at the individual level: medical attention, food, and other assistance must be provided right away. However, this response should be intended to give way gradually to progressively longer-term forms of aid that enable the society to function at progressively larger levels of organization.

Also, even in emergency conditions, some relaxation of the constraints on the proper use of aid should be carefully considered. If we want to maintain the social structures that already exist in times of aid, we should keep in mind that constructive aid does not have to suppress commercial activity. In cases where we want to foster new social structures, we should not stifle entrepreneurial activities from taking root. This does not mean that the development community should stand idly by under looting conditions or foster ill will among recipients of aid, but it does mean that some exploitation on the part of food aid recipients may be allowed to exist. This, after all, is what happens in a developed society, it is the mechanism of food distribution. Only when people find themselves in a competition for resources will they feel the need to cooperate, and only then will they be able to launch activities that will retain or expand their local economies. This is the evolutionary process that leads individuals to aggregate into groups and larger structures—and this is how food aid programs can coexist with a rehabilitation of local infrastructures.

Conversely, development agencies that focus on country projects, need to consider the many finer scales of social organization below that of the country as a whole. Some development agencies have recognized this issue. There is an increasing emphasis in development economics on "microloans:" small, targeted loans that are designed to help individuals with entrepreneurial ideas rather than entire societies at once. This is a start, but like the food aid example, it often errs too far towards aiding the individual. From a complex systems perspective, there are many alternative choices that are better suited to development of social organization at multiple levels of structure. If the goal is development assistance that promotes internal dependencies rather than individual well-being, then we should support development of local communities.

This is where multilevel cooperation and competition comes in. We cannot expect that social organization will arise only from cooperation or only from competition. Instead, recognizing that cooperation and competition are the yin-yang of organization as explained in Part I of the book, we should develop programs that promote competition and cooperation. This might take the form of direct economic competitions, or seemingly secondary social competitions (as in sports). The development of social structures that can lead to effective social systems will only arise from some form of competition at one level and cooperation at another level. Even when aid is provided to solve a local societal problem, it is also a competition for resources between alternative groups that would implement the solution. When the competition occurs between aid organizations, then it is the aid organizations that are gaining capabilities. Having native groups solve the problems creates the local capabilities of organization and action that are themselves a necessary part of development.

A successful strategy for development, then, will involve finding local problems at the community level that need addressing and providing aid to local organizations to combat them. The accumulation of smaller projects that create change at the local level will result in a large effect on a higher level, but without the destabilizing effect that comes with applying a single large force (like building a gigantic dam, or giving a few people decision making authority about allocation of a large amount of funds). An added advantage of working with local projects is that it allows many different approaches to a problem to be tried at once, in different places, to see what works. Successful local initiatives can be used as models for projects with a wider scope (or more similar projects of a small scope), which can attract bigger investments of time and capital.

This approach to development from the local structure upwards is supported by an example from biology: the evolutionary development of organisms, which produce multiple offspring. In a given generation the successful ones multiply, and the unsuccessful ones perish, eventually resulting in a generation more suited to the environment at the moment. We can think of the organisms as analogs of the development projects. When we try many different projects in different places, the successful projects can be copied. Changes in how they are done the next time (variation) leads to continued improvement over time.

Contrast this to the development of a fetus, which takes place in a completely protected environment. The womb is shielded from external problems so that development from a single cell can progress to the point where the entire organism is functional and effectively designed for the demands of the external world. This is sort of like the scaffolding model for development aid, and we've discussed why it's not likely to succeed.

Why must international development follow the model of organism evolution rather than fetal development? The answer is that unlike fetal development, developing countries have not yet settled, through many iterations, on a standard method of growing up. We do not understand social development sufficiently to devise a shielded strategy for development. We are still trying multiple paths to success; this is the hallmark of the evolutionary process. It is the true alternative to the planning trap—small and local plans rather than large and global plans. If we want to try the fetal development strategy eventually, it will take much more experience and a full commitment to understanding how complex systems arise. In the meantime, multiple smaller scale interventions are the best way to identify what will work.

Summary

The subject of international development provides an opportunity for clear discussion of many of the problems that arise with the central design and planning of complex systems. Perhaps this is because the objective is to create functioning societies, the most complex entity we know of. There are three key points that may be used to improve development efforts in the future.

The first point describes the importance of recognizing multiple levels of organizational structure. Traditional efforts to provide assistance tend to focus on either the smallest level of organization, the individual, or the largest, the nation as a whole. Directly helping individuals or directly helping nations results in a weakening of the intermediate levels of structure that are essential to the functioning of a complex society. These intermediate levels are the interactions between people, and groups of people, that comprise trade and commerce, cooperation and competition, which are the basis of economic and social activity. Recognizing the multi-level structure of a social system, and the interplay between these levels, is a first step toward a more effective approach.

The second point describes several aspects of system-environment interactions. In order to perform directed interventions that promote development, careful consideration must be given to the environment, natural and human, in which a target nation or group exists. The existing problems in the Third World are at least in part due to the severe physical environments in tropical climates that have prevented development from occurring naturally. Eventually, a developing nation must today become part of the global economy with all of its tremendous uncertainties and complexities. The dual difficulties facing developing nations coping with both a complex physical and social environment should be recognized. Identifying the external forces that promote or discourage development in each case is an essential second step to understanding development.

The third point discusses the interdependence that arises as a result of intervention. Intervention, particularly strong intervention designed to cause change, generally leads to some form of interdependence between the intervener and the target nation. Recognizing the existence of such entanglements should guide programs to adopt interventions that can, over time, become an integral part of the functioning of the global system.

PRELUDE:

ENLIGHTENED EVOLUTIONARY ENGINEERING

In 1999, without any precursor, I was invited to speak at MITRE to give a one day course on complex systems. MITRE is an organization that provides advice to the government about engineering so that the government can plan and purchase technologically sophisticated systems. There are several parts of MITRE working with branches of the military, as well as the Federal Aviation Administration (FAA), and recently the IRS. In preparing for this program, I called some colleagues at MITRE and at MIT to find out what MITRE was particularly interested in, so that I could provide a specific example in which some of the ideas of complex systems would be relevant. The answer was quite clear: the issue at the center of their awareness was modernization of the air traffic control system.

I remembered that as a graduate student in the early 1980s at MIT, I was on a bus in Cambridge when I asked the person next to me what he was working on. He told me that he was designing the next air traffic control system. I asked him how they were doing that. "From scratch," he responded. This incident has remained clearly in my mind since then. I remember thinking to myself that it would never work. Years later, when asked to prepare the presentation to MITRE, I learned that the same project had been abandoned after costing $3–6 billion and without resulting in a single change being made to the existing remarkably outdated system. Quite aside from the amusement at my youthful prediction being true, I was very pleased that my study of complex systems provided a basis for analysis as well as an approach to successfully performing such large projects. Many people have insights that are consistent with complex systems research. The insights arise from our interactions with the complex world. Still, having a science that can organize these thoughts gives us new opportunities to recognize when one idea is right and another is wrong. Being able to predict ahead of time which projects are possible and which are not, would save governments and corporations a great deal of time, effort and money.

My presentation to MITRE was quite provocative, challenging basic assumptions about what engineering should be in order to be effective. Still, to my surprise, it resonated strongly with many of the participants. They immediately invited me back for more sessions. Because of various constraints, it took a couple of years before a new venue was found for my continued involvement with MITRE, an interaction that continues today.

There is a dual nature to engineering. Engineers are responsible for careful quantitative evaluation of how to achieve objectives, what to do to achieve them, and even (a task that most people find almost impossible) how long it will take to do the task. The other side of engineering is an independent creative "cowboy"-type attitude characteristic of people breaking out of the mold, coming up with novel ideas, implementing them, and changing the world through new technology. This is the culture of high-tech innovation.

When it comes to large highly expensive engineering projects, people who ask engineers to perform them expect a careful failure-proof, well planned out way of executing the project with a guaranteed outcome. In a complex world this is self-defeating. The process of ensuring success is likely to be the process of ensuring failure precisely because the success of projects rests on innovative ideas and solutions to problems that will arise from the creativity of the people involved, not from the plans that can be created in advance.

MITRE originated at about the same time as systems engineering became a field of work and of study. A historical overview of the origins of MITRE can be found in *Rescuing Prometheus* by T. P. Hughes,[68] which describes how MITRE developed out of research projects performed at MIT. The basic idea of systems engineering is that it is possible to take a large and highly complex system that one wants to build, separate it into key parts, give the parts to different groups of people to work on, and coordinate their development so that they can be put together at the end of the process. This mechanism is designed to be applied recursively, so that we separate the large system into parts, then the parts into smaller parts, until each part is small enough for one person to execute. Then we put all of the parts together until the entire system works. This approach seems eminently reasonable. The problem is that it requires someone to coordinate the parts. This problem is entirely analogous to the problem of hierarchical control that was discussed in Part I. When the amount of coordination between the parts is high, the amount of information that one person can know is insufficient for the task.

The real problem is that the method works well, but only up to a point. It works for cases where the system is more complex than a very simple project, and it works for slightly more complex projects if we work harder on the coordination task. It works with some probability for even more complex projects, if we work even harder. Notice that we can work harder and harder, becoming better and better at doing this kind of systems engineering in order to do more complex projects. The problem is that this process has a limit. The limit is given by the amount of time we are willing to wait for the project to be done (if it takes too long we probably don't need that project anymore), and by the ability of a single person to understand or communicate in a given amount of time. What this means is that people will become very good at doing systems engineering and eventually when projects become too complex, they will fail at performing them, even though they are remarkably competent and capable. Aside from the remarkable frustration that this would bring, it would also result in people trying even harder, rather than recognizing that the entire approach no longer works for such problems. Why don't they see this? The answer is that the existing approach worked. It worked on major projects that were important in the development of engineering and defense projects in this country. Today, however, with enough major efforts not succeeding, people recognize that the existing approach is not working.

Since the original lectures at MITRE, I have presented the same material to many other engineering audiences, including recent additional programs at MITRE, military engineering audiences, Boeing, Lockheed-Martin and other military contractors. These presentations have been well received. Over the past two years, my involvement at MITRE has broadened toward developing programs that work rather than just recognizing and identifying what doesn't. It is not enough to identify the overall strategy; one must develop a clear understanding of how to proceed. This is what we are doing today. Much of this effort has been in changing the process of modernization of the Air Operations Center (AOC) in collaboration with

Doug Norman and Mike Kuras. An AOC controls the real time operations of the Air Force. These discussions as well as additional educational efforts are influencing many of the major programs at MITRE.

Chapter 15

Enlightened Evolutionary Engineering[69]

Introduction: Evolutionary Engineering

Systems engineering and management share the common objective of making effective complex systems. Traditionally, there was an important distinction in that engineers built systems consisting of hardware or software or combinations of both, while managers created systems composed of people. Today the distinction is not as great as it used to be, and for many of the complex systems being created today the distinction should not be considered to exist at all. There are two main reasons for this. First, almost any system consists of both people and equipment. Second, the project that creates the system involves both people and equipment and therefore is just as much a management project as an engineering project.[70]

In previous chapters we have discussed some of the organizational failures that have reached the point of crisis in key social systems. In this chapter, we will discuss parallels in major engineering projects. It would seem that the artificial nature of engineered systems would allow them to be created using conventional planning and implementation. However, complex engineered systems, like complex social systems cannot be created in this way. The history of engineering projects over the past two decades is consistent with this understanding. In this chapter we will begin by describing the failures, and then take an additional step toward describing how we can successfully create complex systems.

The main objective of this book is to provide an understanding of how organizations can transform themselves to better suit the complexity of their tasks and environments. We've talked a lot about what the most effective structures for a set of important tasks would look like. However, we have not systematically discussed how to arrive at these structures. Unfortunately, how to arrive at the optimal structure is not obvious. There are many specific details that are important for the overall function of a complex system. In fact, when a system becomes complex enough it becomes impossible for a person to figure it out completely.

If we can't figure it out, then how can we design, manage, control or fix it? How can we fix or improve the highly complex health care system, the education system, or the military? The answer ultimately has to involve evolution because it is the only process that we know of that creates highly complex systems.

In this chapter we'll discuss how to transform an organization into a much more effective structure using an evolutionary approach. Many of the ideas we'll talk about here have been alluded to in previous chapters, but here they'll finally take shape into a comprehensive strategy: enlightened evolutionary engineering. Major engineering projects form the focus of this chapter, but its lessons apply to any complex system you're trying to build or improve dramatically.

Systems engineering successes: the Manhattan and space projects

The two most widely known successful engineering projects are the Manhattan project (which took less than three years to produce a working atomic bomb) and the U.S. space program (which dominated the development of human space flight after Sputnik and assured American leadership in technology for decades). The legacy of these programs is a paradigm for engineering that still carries a great deal of weight today. There are several assumptions inherent in this paradigm. First, it is taken for granted that substantially new technology will be used. Second, this new technology is to be based upon a clear understanding of the basic principles or equations that govern the system (i.e., the relationship between energy and mass, $E = mc^2$, for the Manhattan project, or Newton's laws of mechanics, $F = ma$, and gravitation, $F = -GMm/r^2$, for the space program). Third, the goal of the project and its more specific objectives and specifications can and will be clearly understood. Fourth, based upon these specifications, a design will be created essentially from scratch, this design will be implemented and, consequently the mission will be accomplished.

Large engineering projects today generally continue to follow this paradigm. Projects are driven by a need to replace old "obsolete" systems with new systems, and particularly to use new technology. The time line of the project involves a sequence of stages: a planning stage at the beginning, which gives way successively to a specification stage, a design stage, and an implementation stage. Each of the various stages of the process assumes that managers know exactly what needs to be done and that this information can be included in a specification. Managers are deemed successful or not depending on whether this specification is achieved. On the technical side, modern engineering projects generally involve the integration of systems to create larger systems, their goals include adding multiple functions that have not been possible before, and they are expected to satisfy additional constraints, especially constraints of reliability, safety and security.

Engineering project failures

The images of success in the Manhattan and Space Projects remain with us, but what really happens with major engineering projects is much less satisfactory. Many projects fail and are abandoned despite tremendous investments of time and money. A collection of such project failures is shown in Table 1, with costs ranging from roughly $50 million to $5 billion. The final project in the list, an automation project for dispatching of London ambulances may have cost 20 lives over the 48 disastrous hours during which it was in effect. Each of these projects represents a substantial investment and would not have been abandoned without good reasons.

The most expensive single project that is documented here is the Federal Aviation Administration's (FAA) Advanced Automation System (AAS), the U.S. government's effort to modernize the air traffic control system in the United States. Over the past few decades many of the major difficulties with flight delays and other limitations have been blamed on the technologically antiquated air traffic control system. This system was originally built in the 1950s using equipment based upon vacuum tubes. In the 1960s, mainframe computers were added.[71] By the late 1970s this technology was remarkably obsolete, with functional limitations that would compel any modern engineer into laughter. Yet despite all of this, a modernization effort that cost $3–6 billion from 1982 to 1994 was abandoned without adopting any improvements to the system. Vacuum tubes were still being used! How could a 12-year project costing $3–6 billion not succeed to

Table I: List of Large Scale Engineering Project Failures [72]

System Function – Responsible Organization	**Years of Work (outcome)**	**Approximate Cost** M=Million, B=Billion
Vehicle Registration, Drivers license – California Dept. of Motor Vehicles	1987-1994 (scrapped)	$44M
Automated reservations, ticketing, flight scheduling, fuel delivery, kitchens and general administration – United Air Lines	Late 1960s–Early 1970s (scrapped)	$50M
State wide Automated Child Support System (SACSS) – California	1991-1997 (scrapped)	$110M
Hotel reservations and flights – Hilton, Marriott, Budget, American Airlines	1988-1992 (scrapped)	$125M
Advanced Logistics System – Air Force	1968-1975 (scrapped)	$250M
Taurus Share trading system – British Stock Exchange	1990-1993 (scrapped)	$100–$600M
IRS Tax Systems Moderniza-tion projects	1989-1997 (scrapped)	$4B
FAA Advanced Automation System	**1982-1994 (scrapped)**	**$3–$6B**
London Ambulance Service Computer Aided Dispatch System	1991-1992 (scrapped)	$2.5M, 20 lives

modernize one of the most antiquated systems still in use? The AAS will be a key case study in this chapter because it encapsulates the many things that can go wrong when people try to change complex systems without understanding them.

When a large project like the redesign of the air traffic control system fails, participants and observers often point to a variety of reasons for the failure. In this case, there are several reasons for failure that appear unique. Some have pointed to the U.S. Government procurement process, which involved both the FAA and Congress. Others have argued that the specifications and requirements for the system were never really known. Another possibility is the unrealistic decision to plan a "Big Bang" change that would change the system from the old to the new over a very short time. Yet another candidate for the lack of progress is the emphasis on changing from manual to automated systems. Finally, many blame the ultimate failure on the "safety veto" exercised by air traffic controllers, who could refuse any changes because of their concerns about safety. The latter indeed appears to have been a daunting challenge because the air traffic control system affects the safety of airplanes full of passengers, thus any system failure is likely to lead to the injury or death of many people.

There are convincing reasons to attribute the failure of the AAS to any or all of these problems. Among the actions that have been taken to alleviate these problems, the Information Technology Management Reform Act (IT-MRA), part of the Clinger-Cohen Act of 1996, was created to bring private sector procurement strategies into the government sector in response to evidence of large-scale waste throughout government projects.[73] However, studies of large information technology projects, in the private sector as well as the government, have shown that a significant number of such projects, are completely abandoned after staggering investments of time and money.[74] According to one such major study performed in the mid 1990s, 30% of the projects they surveyed were completely abandoned, an additional 50% of the projects ended up over budget (typically by a factor of two), over schedule by a factor of two, and only meeting a third of the original functional specifications. Ascribing unique reasons to each case may, therefore, not be as constructive as it seems. The high percentage of failures and the remarkable percentage of projects that do not come anywhere near to their specifications suggest that there is a fundamental reason for the difficulty involved in major engineering projects.

Indeed, despite ITMRA and related improvements, successors of the Advanced Automation System that are being developed today have found the process slow and progress limited.[75] From 1995 until 2000, major achievements included replacing mainframe computers, communications switching system devices, and the en-route controller stations. The new equipment continues to be used in a manner following original protocols used for the old equipment. The replacement of the Automated Radar Terminal System at Terminal Radar Facilities responsible for air traffic control near airports, met with greater difficulty. The program to replace these terminals, the Standard Terminal Automation Replacement System (STARS) program, faced many of the problems that affected the Advanced Automation System: cost overruns, delays, and safety vetoes of implementation. Finally, just recently, in 2002, the implementation of a few systems was forced at relatively small airports by FAA emergency decree, overriding concerns about failures in safety tests. In 2003 it was implemented at a few larger airports even though there were still many "bugs" in the system. The full implementation at other airports is expected to take at least eight more years.[76]

A fundamental reason for the difficulties with modern engineering projects is their inherent complexity. The systems that these projects are working with or building have many interdependent parts, so that changes in one part often have effects on other parts of the system. These indirect effects are frequently unanticipated, as are collective behaviors that arise from the mutual interactions of multiple components. Both indirect and collective effects readily cause intolerable failures of the system. Moreover, when the task of the system is intrinsically complex, anticipating the many possible demands that can be placed upon the system, and designing a system that can respond in all of the necessary ways, is not feasible. This problem appears in the form of inadequate specifications, but the fundamental issue is whether it is even possible to generate adequate specifications for a complex system. Our discussion of complexity in the first part of the book suggests that such a specification would be so long as to be unwritable and unreadable, and therefore it is not possible.

Despite the superficial complexity of the Manhattan Project and the space program, the tasks that they were striving to achieve were relatively simple compared to the problem of air traffic control. The Apollo Program, for example, was centered around missions where the goal was to take one piece of equipment, get it to a particular location (earth orbit, lunar orbit, the surface of the moon, etc.), keep it there for a certain duration, and then usually to return it safely to earth. Safety concerns did make the task more difficult when it included carrying a human being on the spacecraft. However, it is a much more complex task to ensure that the three-dimensional paths of any two planes never intersect, everyday and under various different conditions. The trajectories of the many airplanes taking off and landing in a short period of time lead to many chances for error and the necessary safety constraints impose a remarkably low tolerance for failure. The collapse of a particular project may appear to have a specific cause, but an overly high intrinsic complexity of these systems is a problem common to many of them. A chain always breaks first in one particular link, but if the weight it is required to hold is too high, failure of the chain is guaranteed.

Conventional approaches to complexity in engineering

The complexity of engineering projects has been increasing, but this is not to say that this complexity is new. Engineers and managers are generally aware of the complexity of these projects and have developed systematic techniques that are often useful in addressing it. Notions like modularity, abstraction, hierarchy and layering allow engineers to usefully analyze the complex systems they are working with. At a certain level of interdependence, though, these standard approaches become ineffective.

Modularity, an approach that separates a large system into simpler parts that are individually designed and operated, incorrectly assumes that complex system behavior can essentially be reduced to the sum of its parts. A planned decomposition of a system into modules works well for systems that are not too complex. For an automobile, the fuel system and the ignition system can generally be built independently and then put together. However, as systems become more complex, this approach forces engineers to devote increasing attention to designing the interfaces between parts, eventually causing the process to break down.

Engineers use abstraction to simplify the description or specification of the system, extracting the properties of the system they find most relevant and ignoring other details. While this is a useful tool, it assumes that the details that will be provided to one part of the system (module) can be designed independently of details in other parts.

Modularity and abstraction are generalized by various forms of hierarchical and layered specification, whether through the structure of the system, or through the attributes of parts of a system (e.g. in object oriented programming). These two approaches incorrectly portray performance and behavioral relationships between the system parts, often assuming that details can be provided at a later stage of the project.

The same problems plague the mechanisms that managers have developed to organize and coordinate multiple teams of people working on the project. The more the parts being developed depend on each other, the more teams of people must interact. Coordinating the management of the teams, therefore, becomes increasingly difficult.

These mechanisms and techniques of systems engineering are hard to get right, but there's more to the problem than mere difficulty. There are two theorems about complex systems that underlie our analysis of engineering complex systems. The first is the Law of Requisite Variety, which essentially gives a quantitative relationship between the complexity of an engineered system and the complexity of the task it is required to perform.[77] The second is a theorem about functional complexity, which proves that for all practical purposes adequate functional testing of complex engineered systems is impossible.[78] There are simply too many conditions in which they have to operate correctly for testing to be effective. Given such proofs that the problems with engineering complex systems are not due to failure of the systems engineers but a failure of the strategy, we must identify new approaches.

Simplifying objectives

The field of complex systems provides two answers to failures of engineering projects.[79] The first is to simplify objectives when possible. Recognizing that complexity is a crucial property of engineering problems should lead planners to limit as much as possible the complexity of objectives. An estimate of the minimum complexity required to meet the necessary function should be part of the initial process of evaluating an engineering project.

It is easy to imagine two kinds of initial planning meetings when designing a new project. In the first type, everybody brainstorms a wish list of items to be included in the system. This wish list becomes the basis for planning. In the second type, the focus is the minimum set of capabilities that will significantly improve the current situation. The former is surely a prescription for overly complex and unrealistic requirements. The latter has a much greater chance of becoming reality.

The motivation for simplicity may seem obvious, but the tendency to a "wish list" approach to project scoping is often driven by a separation between those that use the system and those that develop them. This is particularly true when multiple vendors are competing for contracts to provide the engineered system. It is still true when in-house developers must justify budgets in competition with other corporate priorities. Creating exaggerated or unrealistic expectations has very limited consequences for those who are involved at the time, as the project will typically be completed years from when the project is approved. Also, it takes quite a bit of effort and understanding to recognize what gives rise to complexity. Not surprisingly, many of the key aspects of modern projects are precisely the ones that add complexity: integration of previously separate systems, multiplicity of functions, and multiple constraints (especially safety constraints). Each of these increases the number of possibilities that the system must encounter and reduces the options that are successful.

In a broader context, we can recognize the approach of limiting complexity in corporate management. It has become an integral part of modern corporate behavior, where there are general trends toward "outsourcing" and focusing on "core competencies." These are clearly ways of reducing the complexity of an organization. They are made possible by the underlying nature of a service economy.

Simplifying the function of an engineered system is not always possible because the necessary or desired core function is itself highly complex, meaning we simply cannot choose to avoid it. We cannot, for example, estimate the complexity necessary for a modernized air traffic control system and then reduce that complexity by deciding that there will be one quarter the number of flights that currently take place. It is even more unthinkable that we could relax the safety constraints for this system. However, this kind of project has already proven impractical for conventional engineering processes.

Enlightened Evolutionary Engineering

In cases like this an evolutionary approach is necessary. The development of evolutionary processes in engineering requires a basic rethinking of how conventional engineering steps are to be accomplished. Since evolution is not a simple process, effective evolutionary strategies must be carefully considered.

Operationally, the key to the creation of an evolutionary process is an agreement to compete and cooperate at different levels of organization. The largest level is the cooperation of the entire set of competitors. The competitors are formed of teams of individuals (and their equipment) that are engaged in operational tasks in competition with each other. The agreement between them consists of cooperatively creating the environment to provide the infrastructure and rules for the competition.

The basic concept of designing an evolutionary process is to create an environment that fosters a continuous process of innovation in the system itself. Think of the individual parts of the system—hardware, software or people involved in executing tasks as part of the system—as analogous to biological organisms in a natural environment. In an evolutionary process changes to these parts will take place through substitutions that might involve new designs, training, or changes in how they are arranged or work together. This replacement of components involves changes in one part of the system, not in every similar part of the system. Any one of the individual changes of one of the software or hardware elements could be performed as a conventional engineering process if it is not too complex. However, even when the same component exists in many parts of the system, changes are not imposed on all of these parts at the same time. Multiple teams are involved in design and implementation of these changes. This is the *opposite* of standardization—instead of imposing uniformity, we are explicitly imposing variety on the system.

The development environment should be constructed so that exploration of possibilities can be accomplished in a rapid manner. If experience with a particular changed component indicates improved performance, then that component may be more widely adopted by the individuals involved. This is a kind of evolution that happens through informed selection. The process of selection explicitly entails feedback about aggregate system performance in the context of real world tasks.

Thus the process of innovation in the context of large systems engineering projects will involve multiple variants of equipment, software, training or human roles that perform similar tasks in parallel. Let's follow a single piece of equipment through its own process of change, which occurs in several stages. In the first stage a new variant of the equipment is introduced in parallel with the old version. This new equipment can be developed using the conventional development process—the individuals or teams developing it may still use well-known and tested strategies for planning, specification, design, and implementation. Locally, this variant may perform better or worse than others. Overall, however, introducing it does not significantly affect the performance of the entire system, because other older versions of the equipment are operating in parallel. If the new variant is more effective than the older one, then in the second stage others may choose to adopt it in other parts of the system. As adoption occurs there is a load transfer from older versions to the new version in the context of competition, both in the local context and in the larger context of the entire system. In the third stage, the older systems are kept around for longer than they are needed, used for a smaller and smaller part of the load until eventually they are discarded 'naturally,'

In essence, the new pieces of equipment are competing with the older ones for the right to perform the necessary tasks. When they do so successfully, they will tend to be adopted and the older ones will be phased out. However, following a single process of innovation will produce a biased view of the evolutionary engineering process. Instead, the key is recognizing the *variety* of possibilities and subsystems that exist at any one time and how they act together in the process of innovation. The variety arises from having multiple new innovations in components introduced in different parts of the system and then allowing them to compete to increase their usage proportion.

The conventional development process currently used in major engineering projects is not entirely abandoned in the evolutionary context. Instead, it is placed within the larger context of the evolutionary process. What is different is that new alternative components are introduced in parallel, which ensures redundancy and robustness. At the same time, the ongoing variety provides robustness to changes in the function of the system. If the function of the system is suddenly changed, the system can adapt rapidly because there are various possible variants of subsystems that can be employed.

The different components that are introduced should be developed by different design teams that work separately on their designs. The existence of multiple small teams designing new components that are to be introduced in parallel, and the separation between these teams as they design components, ensures maximum innovation in the same way that the barriers that divide subdivided networks (see Chapter 3) ensure creativity.

The usual distinction between human beings and equipment is not relevant to the way we should think about the evolutionary engineering process. We include both human beings and machines (computers, communication devices, electronic networks, etc) as parts of the system. Changes in training and in how people interact are modifications of the system, just as changes in the equipment are modifications of the system. People and equipment are both part of the system actions. They are also both engaged in the process of system modification, because design teams involve people and their equipment. We can consider the process of creating system components (training, design, engineering, construction) also to be part of the system activity itself. Therefore, quite generally, human beings and computers are interactive agents in the process of design, development, and implementation, as well as in the functioning of the system.

Evolution is a process of cyclical feedback and the role of the dynamics of this feedback often leads to a need to balance different performance aspects that are mutually contradictory. The central contradiction is that after some period of time, the process of selection and competition generally gives rise to a single dominant type that inhibits innovation. This is known as the "founder effect" in biology and sociology and as monopolization in economics. To avoid internal inhibition of change, the process must be designed to promote change and destabilize uniform solutions to problems, when it is appropriate (i.e., dictated by system performance in the context of interaction and feedback with the external environment). One way to do this is to adopt an analogue from biology—the generation time or lifetime of the organism—and require a certain rate at which new innovations are introduced. Such promotions of change might appear counter to the process of selection itself, since over the short term, promoting alternatives to established solutions appears to be counter to selection of the most effective system known at that time.

Another balance that must be reached is between promoting the propagation and adoption of improved systems and inhibiting propagation in order to allow sufficient time for testing. If adoption is too rapid, a solution that appears effective over the short term may come to dominate before it is tested in circumstances that are rare but important, leading to large scale failure when these circumstances arise.[80] If adoption is too slow, the system cannot effectively evolve, giving rise to an inhibition of change as previously noted.

Understanding the balance needed is a current area of research and simple guidelines are not yet known. The best that can be done is to alert the manager of the evolutionary engineering process to the symptoms of effective and ineffective evolutionary change so that they can be recognized and modifications "on the fly" can be made in the evolutionary environment with the objective of improving the balance. Since the evolutionary engineering process will be designed in such a way that iterative refinement of the process itself is possible, this is not a critical limitation. Indeed, this is consistent with the idea that comprehensive advance planning (as currently understood) is often not possible and that the system is designed to be effective in an adaptive process.

Application to air traffic control

How can we apply evolutionary processes to implement change in a context where risk of catastrophe is high? Our primary example will be the air traffic control system discussed earlier. Similar problems exist in other high-risk contexts including the nuclear power industry, and in the military.

The problem with innovation in the air traffic control system has not been solved because we still have the "safety veto:" How can we introduce changes in what an air traffic controller is doing without introducing grave risks to people in airplanes? This was the problem that eventually derailed the Advanced Automation System. Even today, the process of innovation in the air traffic control system is very slow because of a need to extensively test any proposed change.

What is important to realize is that there's actually an existing process of innovation and introduction of new components into the air traffic control system that has been operating smoothly for decades: the training of new air traffic controllers. Air traffic controllers undergo extensive preparation, as well as multi-stage on-the-job training. Consider the stage in which the air traffic controller in training is acting as Controller, but a second Controller (supervisor) is present and has override capability over the trainee. Thus, when a person is being trained, he or she performs the task under supervision and the supervisor's override privilege prevents accidents from happening.

This same mechanism can be used for air traffic control innovation in hardware and software. The key is to have two different stations that can perform the same functions, where one of them has an innovation in hardware or software and the other retains the more conventional system, while having override capability over the first.[81] In this case, both of the human air traffic controllers would be experienced controllers, not trainees. This dual system can be used to test new options for air traffic control stations, while providing the same standard of safety. (This dual system is not the same as the current dual system of Radar Controller and Radar Associate Controller, but is either in addition to, or possibly as a substantial modification of, this system.)

There are many possible technological innovations that could be tested in this way. For example, the traditional air traffic control stations consist of monochrome screens with visual sweeps of the air space. Any change in this system could introduce problems. The sweeping of the screen appears obsolete compared to modern screen technology and only a residue of the limited technology that existed in the 1950s. However, a process of sweeping may be useful to keep a person alert in the context of continuous monitoring. In this case, an unchanging screen may lead to failures rather than improvements. Similarly, adding color might seem a good idea, but the color might be distracting if the way it is done attracts the attention of air traffic controllers to unimportant information.

How can these questions be tested safely? They can be tested by introducing in a trainer context a version of new screens that have continuous displays, color displays, or other changes. Allowing sufficient time for an air traffic controller to become used to the new system, the override capability can be retained for an extended period of time to test the system under many contexts: day, night, low and high traffic, extreme weather, etc. Such redundant execution of tasks is needed, as well as maintaining older solutions that are more extensively tested. Indeed, we can expect that many variations on displays would be distracting or ineffective at bringing the key information to the attention of the air traffic controllers. Without such extensive field testing mistakes would surely be made.

The idea of using a double "trainer" has a biological analogy: the double set of chromosomes in animals. The double set acts at least in part as a security system to buffer the effects of changes in the genome. In this case, either of the chromosomes may be changed so that there are two different parallel systems that are both undergoing change. The probability of failure would be high, except that they both exist and failure of one does not generally lead to failure of function of the organism.

A system using this double trainer method would ideally have many if not all air traffic controllers working in pairs, where one has override capability. It is also possible to set up a double override capability to allow mutual oversight. It may be argued that the cost of having double the number of air traffic controllers is prohibitive. However, the alternative has already been demonstrated to be ineffective at the level of $3–6 billion in direct wasted expenses for modernization. This doesn't even include the ongoing annual losses due to canceled and delayed flights caused by ineffectiveness of the air traffic control system. After all, it is these costs that motivated the spending of billions on improvement efforts.

Redundancy is a general mechanism for achieving reliability and security in function. We've discussed this quite a bit in the context of medical errors, another area with crucial safety constraints. The level of redundancy required increases as the level of safety required increases.

The importance of redundant execution of tasks can be understood in the context of the air traffic control system. The air traffic control system exists at the maximum level of functionality. In this context safety problems are highly probable when any change is introduced into the system. By introducing redundancy, an additional level of safety is introduced. Once there is additional safety in the system through redundancy, there can be a possibility of change in the system. Even though each change that is introduced is small, rapid change can result because of the parallel testing of small changes at many different locations.

Note that in this process the people making the decisions about what changes to make through the process of wider adoption are the people who are closest to the process itself, in this case the air traffic controllers. In conventional engineering, the people making most of the decisions about changes are far away from the execution process and often do not have direct experience with it (or at least, *recent* direct experience). At the same time, the people who are introducing the innovations in technology remain the people who are most familiar with it, the engineers and designers of systems that are then tested and adopted by real world evaluation.

According to conventional engineering methods, once the overall concept, objectives or functionality of the system desired is determined, the role of engineering management is to provide a sequence of progressively more detailed specifications of the system (i.e. the "waterfall method"). In the context of evolutionary engineering, the role of management becomes much more indirect. Rather than specifying the system, management simply specifies a process and context for the development of the system. Goals for the system (as specified by the desired functionality) are embedded in the context of the tasks involved in this process. For example, the process could involve the operation of double air traffic control stations, while the functional goals are implicitly embodied through the direct evaluation of functional capabilities.

This kind of indirect management may seem to be almost superfluous. It is, however, essential in order to enable the process to occur and be effective. Ultimately, one of the most important roles of management in this approach is to establish mechanisms by which hidden consequences of changes are made more visible. They may be hidden because the consequences are longer-term or larger-scale or cumulative. For example, in the case of the air traffic control system, one key to effective imposition of safety is the availability of direct measures of proximity to failure, measures of "near misses." When changes are implemented in the system, direct measures of near misses provide feedback about the effectiveness of the change in the context of the system. This feedback can then be used to determine when a particular innovation should be more widely adopted.

Designing rules of the game

To promote effective adoption of the evolutionary engineering model it is important to anchor it in common experience. We've developed the analogy more extensively in the beginning of this book, but it is useful to note that the most common experience we have with evolutionary analogues is in organized games and sports. The framework of the game in this case is that the immediate goal is successful completion of tasks, just like the goal in biology is survival and reproduction through successful consumption of resources. The agents of the system—human beings, hardware and software—are competing for the right to perform tasks. We can also think about this as an economy or market in which performing the tasks is the objective. In professional sports and economics there are extrinsic financial rewards for effective execution. The evolutionary process suggests, however, that success can be rewarded by replication, which in this context is wider adoption of innovations. Indeed, the competitive spirit of human beings leads to a preference that the innovations that they contribute to or are using will be more widely adopted. Thus, the possibility of wider adoption should be sufficient to create a dynamic of mutual influence and constructive competition. Management can constructively foster competitive sportsmanship between individuals and especially between teams.

In keeping with the sports analogy, it makes intuitive sense to think about creating the evolutionary engineering context as setting up the "rules of the game." These rules should themselves be simple. For the design of highly complex systems, the complexity of tasks to be performed is the source of functional complexity of demands on the system. Thus the objective of designing the rules of the game should be to avoid additional complexity due to the rules themselves. Only the rules that are truly necessary should be established; these rules should be as simple as possible.

It is crucial that when managers set the rules of the game, they *not* specify the actual mechanism or structure of the engineering solutions of the problem. Instead, the managers should expect that a diversity of unforeseen possible solutions of different aspects of the problem will be adopted eventually. Limitations on the diversity of possibilities should be avoided, unless they are important to how the game is played. The system can then end up with multiple types of parts of various sizes and capabilities. How strongly-integrated or weakly-coordinated the parts are will be determined through the evolutionary engineering process; if strongly-integrated parts are more successful at the tasks, then they'll most likely end up adopted. What is essential is that the parts are usable in the field; integration of the parts into collectives is not the objective at all. In fact, the more closely coupled parts are, the more difficult change will be. Thus, in the competition between evolving parts, the rate at which innovation can take place, the "evolvability" of the system, is higher when there are smaller parts. As a matter of guidance, larger scale integrated systems should only be used when smaller more loosely coordinated parts cannot perform the necessary functions.

Artificial evolution beyond the natural evolutionary model

Enlightened evolutionary engineering provides an important paradigm for improving the effectiveness of major engineering projects. While a discussion of lessons from natural evolution provides a basis for this discussion, there are at least two contexts where we can find examples of "artificial' evolutionary processes that are specifically designed to accelerate the evolutionary process in order to achieve adaptation at a rapid rate. These are found in the immune system and in the process of learning, which have both been introduced earlier in this book.

The process of immune system "maturation," by which the immune system improves its ability to fight alien substances (antigens), involves the evolutionary change of agents that are part of the immune response. This occurs by replication of molecules, "antibodies," that are then selected through their effectiveness in binding (affinity) to thc antigens. In human beings, as well as other mammals, the process of replication and selection is accelerated in special places called germinal centers. In these centers, fragments of antigens are stored and used to test the affinity of antigens produced by an accelerated process of evolutionary change involving high replication rates, a shortened generation time and rapid mutation. These changes and other aspects of the design of germinal centers have been shown to be highly effective at accelerating adaptation.[82] The analogy to an engineering context would be the use of a simulation center where accelerated testing and exploration of prototypes can be performed. The use of some level of simulated context is common for testing engineering projects. The biological analogy suggests incorporation of a multiple iterative parallel evolutionary strategy in simulated and real contexts, with a highly accelerated evolutionary process in the simulated environment.

The process of learning that occurs to train the modular architecture of the brain includes the off-line time of sleeping.[83] It has been proposed[83] that sleep has a key psychofunctional role in the testing and refinement of separated modular components of a modular architecture. This role allows simplification of individual parts, allowing the entire system to learn new functions while avoiding overload of the components. The analogy in an engineering context is exercise, testing and redesign of individual components in a context where the individual component functional role is evaluated while at least partially dissociated from the rest of the system.

These biological examples suggest that off-line experimentation (separated from the "field" in either space or time) can be combined with actual field experimentation. By increasing the amount of experimentation, we accelerate adoption of effective strategies and components. While it is not known if natural evolution creates such off-line opportunities, quasi-artificial evolutionary processes can certainly use them as an integral part of the process.

Conclusions

The complexity of large engineering projects has led to the abandonment of many expensive projects and led to highly impaired implementations in other cases. The cause of such failures is the complexity of the projects themselves. A systematic approach to complex systems development requires an evolutionary strategy where the individuals and the technology (hardware and software) are all part of the evolutionary process. This evolutionary process must itself be designed to enable rapid changes, while ensuring the robustness of the system and overall system safety. The systematic application of evolutionary process in this context is an essential aspect of innovation when complex systems with complex functions and tasks are to be created.

This chapter has proposed that large engineering projects should be managed as evolutionary processes undergoing continuous rapid improvement through adaptive innovation. This innovation occurs through iterative incremental changes performed in parallel and thus is linked to diverse small subsystems of various sizes and relationships. Constraints and dependencies decrease complexity, thus adaptability, and should be imposed only when necessary. The evolutionary context must establish necessary security for task performance and for the system that is performing the tasks. In this context, people and technology are agents that are involved in design, implementation and function. Management's basic oversight (meta) tasks are to create a context and design the process of innovation and to shorten the natural feedback loops through extended measures of performance. The prime directive in the context of the large-scale engineering projects is to simplify whenever possible, avoiding strategies that unnecessarily introduce complexity and impede adaptability.

The same strategy that we have described here in the context of engineering systems can be applied to health care systems and even (somewhat differently) in the education system. In the health care system, the discussion is quite similar to the engineering one. The process of evolution engages teams of medical practitioners and their equipment, just as here we discussed teams of air traffic controllers and their equipment. While the focus in the context of the air traffic control system was on the equipment, the behavior of the air traffic controllers was also part of the system behavior. In the context of the health care system, similarly, improvements in the equipment along with the patterns of behavior of the medical practitioners are all part of the evolutionary process. One way to think about patterns of behavior is as formal protocols. Introducing a new protocol, or other change in how individuals and teams behave, is the elementary innovation. To create such an improvement process, feedback is necessary using measures of performance of tasks in the context of the real world. Also similarly, ideas about how to improve the system must be subject to real world testing before wider adoption. Moreover, ongoing change requires that a high variety of possible solutions be constantly undergoing evaluation to determine what steps will improve the system in a context where complexity is high. This concept of innovation and improvement should be recognized by medical professionals as very close to the traditional way that innovation and improvement of medical care occurs. If a practitioner has an idea about how to make an improvement, he or she may test it in a safe way and if it is successful in a number of cases, others may adopt it. The key modification here is that the feedback today has to reflect the effectiveness of teams rather than individuals. It is generally not easy for an individual to be aware of the effectiveness of teams that he or she is part of, or the reasons for this effectiveness, without additional mechanisms for feedback. Explicitly and effectively setting up evaluation and feedback of team effectiveness is therefore an essential part of management function in fostering the evolutionary process. Fostering this evolutionary process should accelerate improvement dramatically.

The process of evolutionary improvement of the education system also requires ongoing evaluation. In this case, we can at least in part use the judgment of students and parents to perform selection among desired educational environments. A more subtle aspect of the evolution in this context involves how the evolutionary process applies to students. The selection of students for multiple niches, i.e. the recognition that there are many different ways for children to perform effectively, means that the key is to identify which set of educational activities is most suited to a particular child. The evolutionary dynamics is in large part a selection of which specialized environment a child should be in rather than improvement of the system performing the education. Similar issues arise in the context of evolutionary processes in engineering and medical practice when different equipment or behavior patterns are useful under different conditions, leading to an intrinsic need for specialization in how tasks are performed, and a routing of tasks to the appropriate place.

The wide applicability of evolutionary change is a fundamental expression of the unique status it has as the only mechanism we know by which systems that are both effective and highly complex can arise. To solve complex problems we need effective complex systems. Therefore, we can expect that evolution will play an increasing role in our everyday activities.

PRELUDE:

GLOBAL CONTROL, ETHNIC VIOLENCE AND TERRORISM

From the time of the creation of the New England Complex Systems Institute and the first International Conference on Complex Systems in 1997, various members of the intelligence community have expressed interest in learning about complex systems to gain insights relevant to their own concerns. In one of a series of interactions, Mai Nguyen and a colleague from one of the intelligence agencies visited NECSI during the summer of 2001. They were interested in enhancing the ability of the intelligence community to anticipate the locations of ethnic violence. They gave me an article describing a case study of a town in Indonesia that had been the site of terrible violence between Christians and Moslems. They asked me about creating a model that would predict whether a particular town would be the site of such violence, taking into consideration various factors about the town. There are studies that identify particular aspects of a country that are correlated with the rise of violence. Many factors might be considered. Some of these factors might be political, such as the type of government or the behavior of leadership, some might be educational, some financial, and so on. The correlational studies use existing events where violence occurs to look at the factors that seem to be associated with and might help determine the likelihood that ethnic violence will occur at a given location. Another approach, the approach that the visitors expected I would take, would be to identify a set of key causal influences, social, political, economic, historical, and develop a model that would take these causal influences into consideration in describing the reason that one particular town would become the site of violence.

My answer to them was based on a different kind of analysis, one dealing with the overall characteristics of the dynamics of civilization today. These issues were on my mind when I wrote my textbook several years previously. I felt inhibited from discussing them in the textbook because of the sensitive nature of the topic. However, I described them in responding to the question posed about ethnic violence.

The analysis of social change that is provided by a multiscale perspective suggests that over time it is becoming unreasonable to expect all groups of people to mix peacefully. In some cases, there is a natural process of separation that results from this phenomenon. Where separation is taking place, but areas are still mixed, conflict naturally occurs due to the frustration of desires of the different groups for control. A quite reasonable solution to conflict in this case, therefore, is to resolve issues of control peacefully early on rather than waiting for violence to occur. If there are appropriate boundaries between the groups, they may exist peacefully side by side, but without mixing. Thus, adopting the approach of arranging for separation, like the separation of two children who frequently fight, or like the old saying "good fences make good neighbors"[84] seems a good strategy. Recognizing that local wars, often due to ethnic violence, have been estimated to have taken over 40 million lives in the 50 years after the world wars,[85] and with many existing conflicts today and new conflicts arising annually, perhaps we should recognize that insisting that all people live peacefully together in a single mixed community is not necessary, rather all people can live peacefully with appropriate separation.

Viewed globally, the world today appears to be undergoing a natural process of separation between certain groups. The process is similar to the separation between oil and water. This separation acts as a kind of pattern formation, similar also to the kids in kindergarten in Chapter 2 separating into regions of those who wanted particular kinds of toys. The most prominent group that is separating from others is the Islamic world. Changes that are taking place in the rest of the world, and changes that are taking place in the Islamic world are making the two groups less compatible as far as mixed coexistence, requiring more separation for peaceful coexistence. As this process takes place, violence arises in areas where the natural process of separation is not occurring fast enough or smoothly enough to satisfy the people who are mixed at the boundaries. Arranging for peaceful, voluntary separation seems to be the best alternative to the violence that is occurring today in many parts of the world.

In view of this realization, I suggested to the intelligence community visitors to take out a map of the world and mark on it the boundary between Islam and other groups. At locations where this boundary was unclear and populations were mixed, there would be ethnic violence. It is important to emphasize that as far as I am concerned this is a case of global pattern formation and differentiation, not a story of good and evil. The model of separation does not value one side or the other, but recognizes that the boundary between them is a dynamic and often hazardous place to live. The reason for my statement has to do with the dramatically different trends in the Islamic world than the rest of the world over the past few decades as discussed in this chapter. A similar but not quite the same conclusion was reached earlier by Sam Huntington in his book, *The Clash of Civilizations and the Remaking of World Order*.[86] Unlike Huntington, I do not suggest that this is about intrinsic conflict between "civilizations," but rather about the dynamics of domain and boundary formation within a global civilization. Recognizing this suggests a different approach to solving the problem: Clear boundaries.

Creating an effective global society without violence will require a new form of respect and appreciation of cultural differences. This respect for differences occurs at the group rather than at the individual level. It is not enough to consider individual freedoms in establishing choice of culture within a diverse society, it is also necessary to consider the rights of groups to establish collective behaviors that are not the same as those that others would choose. Only by developing this form of respect can we diffuse ethnic violence and conflict, i.e. conflict at the group level.

A couple of months after my discussion with these visitors the events of 9/11 occurred. Today, after 9/11, it is more acceptable to discuss these issues in public. Still, not everyone will agree that my conclusions will be the right course of action. Time will tell.

Chapter 16

Global control, Ethnic Violence and Terrorism

Toward decentralized control

In considering the properties of ethnic violence and terrorism, it is useful to step back and consider some overall societal changes that have been taking place over the past few decades. In the first part of the book, and in other chapters in this part of the book, I discussed the role of hierarchical control in organizations. The conclusion reached was that a hierarchically controlled system is not effective when presented with a highly complex context that requires significant coordination of the collective behaviors of the organization. Historical trends suggest that we have reached a point where the socio-economic environment is too complex for hierarchical control of organizations.

During the 1980s, many countries changed from hierarchical control to more distributed control forms of government. This is apparent in Central and South America where dictatorial forms of government in Argentina, Bolivia, Brazil, Chile, Ecuador, El Salvador, Guatemala, Nicaragua, Panama, Paraguay, Peru, Suriname and Uruguay became more democratic in their political institutions with more open economic systems. It would not have been surprising for any one of these to change because there had been many switches back and forth before that time. What is remarkable is that over a period of ten years, all of them switched in one direction and have stayed that way ever since. The only centrally controlled system remaining in the Western hemisphere is Cuba. Elsewhere in the world there are also examples of such changes, notably in Greece, the Philippines, and South Africa.

The collapse of the Soviet Union at the end of the 1980s and the growth of legal corporate ownership and free markets within communist China over the same decade also reflect dramatic changes away from hierarchically controlled governments. Very few people anticipated the Soviet collapse because it was counter to the experience of history. Governments generally don't give up control or power, even when circumstances are very difficult for the government or for the people of the country. Often the government itself can be responsible for economic and social problems and still persist.

Indeed, what is particularly remarkable about many (not all) of these transitions is that they were peaceful. This is counter to the historical pattern that can be seen in the French revolution at the end of the 18th century or the Russian revolution at the beginning of the 20th century. The French and Russian revolutions began with an effort to reform a government that was not functioning well. Gradually the reform process became more radical, then there was a bloody revolution, which led to a new but still hierarchical form of government. This dynamic, which led back to a hierarchy, suggests that despite the limitations of hierarchical control, it was the stable form of government in the face of social disorder. By contrast, many of the more recent changes in government have been peaceful. In some cases, the individual or individuals in control simply "gave up" this control.

The movement away from hierarchical governments was not the only place where major changes in control occurred. During this same period, changes in corporate control structure took place in many companies in the U.S. and elsewhere. Management change became a major factor starting in the early 1980s with the widespread adoption of Total Quality Management (TQM). The principles of TQM led to changes in the roles of managers. From our perspective, the main point is that teams of individuals become responsible for decisions rather than a single person, e.g., the CEO. In the 1980s and continuing through the 1990s, TQM and other approaches such as the Learning Organization, Reengineering, High Performance Organization, and Lean Manufacturing, have led organizations to adopt structures that are more distributed in control and in which information passes laterally through the organization instead of up and down the hierarchy.

The dramatic changes in control in governments, both dictatorships and communist, and the similarly widespread changes in corporate control suggest that the global environment has become too complex for a single person in charge of a hierarchical organization to respond to. Therefore, centrally controlled, and even decentralized but still hierarchical structures where the large-scale behaviors are centrally controlled became ineffective. This is consistent with the widespread recognition of the complexity of modern life. It is also consistent with the increasing global interdependence that exposes countries and corporations to many and varied forces that require effective response.

More directly, the implication is that the large-scale complexity of human organizations has reached the point where it is greater than that of a single human being at the scale of human communication. The reason we feel this complexity in an intense way is that when the complexity is larger than a human being, it is not only difficult to control, it is also impossible to understand fully. This is why government and corporate leaders have often by themselves made the decision to transfer their control to others. If they could figure out what to do to solve problems, they would not have done so.

We can also take a different approach to seeing the way hierarchical control doesn't work for complex systems. Consider the food supply to a large city, for example, Boston. Think of all the different kinds of food, the different ways food is delivered, trucks, trains, ships, and airplanes. Some of it is refrigerated; much of it has to arrive within a limited time. Think of all the storage facilities that are involved in storing this food. Also, think of all the different places it goes: supermarkets, restaurants and other institutions. The right foods have to arrive at the right time in the right amounts, and so on. What would happen if we tried to control this centrally? The answer is that we would have to limit the number of types of food and the number of places that it arrived; even then things would arrive at the wrong times in the wrong quantities. This scenario is reminiscent of food supply in Moscow before the breakup of the Soviet Union.[87]

In the Soviet Union tremendous effort was devoted to planning the economy. There was a general five-year plan, and then there were detailed one-year plans that were broken up further into one-month plans. They used a form of computerized scientific management, as well as a careful negotiation process between individuals who were responsible for individual enterprises in the system. In the one-year plans, the flow of materials, products, labor and money was directly specified for each product within each enterprise. Not only was what went in and out specified but also where it came from and where it went to. On a daily basis (and then weekly, monthly and yearly) the flows of money were monitored by the banking system so that they corresponded to the plans. The prices were set centrally so that the flows of money corresponded to the flow of materials, products and labor. The planners were well aware of the U.S. free market system and they viewed it as wasteful. Planning, they believed, would lead to increased efficiency due to an elimination of wasteful duplication of effort. In the free market system there are multiple companies doing the same thing. This repetition of effort seems to planners to be a waste of labor and capital.

How well did the planned system work?

In a supermarket in Moscow, the total number of possible foods you might find was only roughly a hundred. Start counting them: sugar, salt, pepper, bread (a few kinds), meat (beef, chicken, pork), milk, cheese (a few kinds), macaroni, potatoes, cabbage, beets, carrots, pickles, and so on. There was almost no fresh fruit and vegetables, though a few were found in a limited season: tomatoes and fresh cucumbers from August to October, plums in September, apples in the fall, and strawberries for two weeks at the beginning of summer. Forget packaging. There was none.

This is not even the whole story. Most of the time even these foods were not available. It was a system where scarcity was the rule. People had to be satisfied with what there was, not what they wanted. They waited in line for food and were alert to food arrivals in stores to be sure to get some. Much of the food was often partly spoiled and beer and milk were often watered down. Waste was very high, 20–50%, even though the items were very scarce. A substantial fraction of fresh fruit rotted in warehouses. Because of the scarcity people couldn't be picky about what they bought. Waiting in line and shopping generally took a substantial fraction of people's time and a significant fraction of income was spent on food.

This was the main food system of the Soviet Union. There were several others that provided the means for people to get additional items. There were farmers' markets, black markets, and some stores that were exclusive to the privileged few. The farmers' markets were the main source of additional food options, though at significantly higher prices.

Contrast this with the U.S. food supply system at the time.[88] American supermarkets in this period were stocked with well over 10,000 products (today nearly 40,000), selected from over a hundred thousand possible products by supermarket owners (with 20,000 new products introduced annually, only a small fraction of which succeed). Many forms of processed and prepared food were readily available. Food of various types, prices and qualities was available at essentially all times (24/7) and in all locations. The economy as a whole was and is consumer-limited rather than supply-limited, so that advertising is necessary for sellers to promote their products.

There is a direct connection between the failure of the Soviet food system to provide adequate improvement and the collapse of the Soviet Union. The person "in charge" of agriculture in the USSR from 1978–1985 was Mikhail Gorbachev, before he rose to become General Secretary in 1985. His college degree was as an agronomist-economist. The ineffectiveness of the agricultural system led to Gorbachev's efforts to change the Soviet system and might be considered among the immediate causes of the collapse of the USSR. The leaders of the USSR were very aware of the comparison of their effectiveness as measured in comparison with the U.S. and other countries. Thus, we would be well justified in saying that the inability to perform the complex task of food production and supply, as compared with the effectiveness in other places, contributed to the downfall of the centrally planned economy of the USSR.

The collapse of the Soviet Union, the free markets in China, the change of many governments from dictatorships to more democratic systems and the implementation of TQM in corporations all point to the inability of central control to effectively manage the complexity of modern social organizations in the face of complex external forces and demands.

Exceptions

Once we recognize the dramatic tendency in much of the world toward decentralized control, it is interesting to consider where the exceptions exist. Two of the most prominent countries that have not followed this trend are Cuba and North Korea. Both of these are small countries that are almost completely isolated from the rest of the world because of the persistence of a conflict with the U.S. This isolation prevents these countries from being exposed to the complexity of the world, a complexity that other countries must cope with. The result is that the internal society remains simple and central control continues to be effective even if it is difficult for the people to tolerate, as manifest in the case of North Korea where the food supply has been severely limited in recent years.

Interestingly, this analysis suggests that the U.S. policies in isolating these countries are themselves responsible for retention of the governments that are an anathema to the U.S. Of course, the reasons for their isolation by the U.S. may have nothing to do with any desire to change their form of government. Political conclusions aside, the existence of central control in these contexts can be understood directly from the issues of environmental complexity that we have analyzed. Simplifying the external environment that these countries operate in, allows their centrally controlled structures to continue.

Toward central control

There are two other parts of the world, however, where central control continues to be widespread. The first is in the Arab, and more broadly the Islamic world, while the second is in sub-Saharan Africa. Understanding the first is central to topic of this chapter. The latter is a highly diverse but generally undeveloped area that is a context for many of the key global problems of poverty, development, ethnic violence, and disease.

When we consider the trends of central control in the Islamic world, we find that many countries have become more centrally controlled rather than less so, over the same time period when dictatorships and communist regimes elsewhere have disappeared. Well-known examples of societies that were much more open before this period than at the end, include Lebanon and Iran. In many cases, religious extremism has been a clear driving force for change toward a closed and restricted society.

Taking the list of all Islamic countries, we find that monarchies tend to be located near the origins of Islam, in the Arabian peninsula. Radiating outward from there we find constitutional monarchies, dictatorships/military strongmen, republics with self-perpetuating authoritarian presidents, and a few democratic republics in the farthest areas, particularly Turkey and Western Africa, and (recently) Indonesia. The trend toward centralization has been clear throughout much of the region. The stability of the centralized governments has become apparent with the passing of control from father to son in Syria and Jordan, and the transfer of power in Egypt. Some recent exceptions that represent a trend toward democratization near the boundaries (Indonesia, Pakistan, Western African states), have yet to demonstrate their stability, with Pakistan already reverting, at least temporarily, to military control.

A list of approximate governmental forms is as follows:[89]

Arabian peninsula:

Bahrain (constitutional monarchy)
Kuwait (monarchy)
Oman (monarchy)
Qatar (monarchy)
Saudi Arabia (monarchy)
Yemen (republic—strong president)
United Arab Emirates (federated kingdoms).

Northwest of Arabia:
Jordan (monarchy)
Lebanon (republic, post civil war)
Syria (military regime/dictatorship)
Turkey (democracy).

Northeast of Arabia:
Afghanistan (theocratic military rule, warring militias [prior to U.S. military action])
Iran (theocratic republic)
Iraq (republic—military strongman [prior to U.S. military action])
Pakistan (military strongman).

Further Northeast—former Soviet Republics (all to be considered in transition):
Azerbaijan (republic)
Turkmenistan (republic—president for life)
Uzbekistan (republic—authoritarian president)
Kyrgyzstan (republic)
Kazakhstan (republic—authoritarian president)
Tajikistan (republic, civil unrest).

East of Arabia including Southeast Asia:
Bangladesh (parliamentary democracy)
Brunei (monarchy)
Comoros (unstable military rule)
Indonesia (military strongman till 1998, republic & ethnic violence since)
Malaysia (constitutional monarchy)
Maldives (republic, same president for 25 years).

South of Arabia (Across the Gulf of Aden):
Djibouti (republic)
Somalia (warlords)

West of Arabia—North Africa (bordering the Mediterranean Sea):
Algeria (republic—strong president)

Egypt (republic—strong president)
Libya (military dictatorship)
Morocco (constitutional monarchy)
Tunisia (republic—one party).
West of Arabia—Next tier Africa (bordering North Africa):
Chad (republic—oligarchic control—conflict with south part)
Niger (republic from 1999)
Sudan (military/Islamic regime—conflict with south part).
Further West—West Africa:
Gambia (republic, from 1996)
Guinea (republic—military ruler still president)
Mali (republic, from 1991)
Mauritania (republic—one party)
Senegal (republic)
Sierra Leone (republic—civil unrest).

Others have made this observation, particularly since 9/11. In an article by Fareed Zakaria in Newsweek[90] this point was explicitly made. He states, "In an almost unthinkable reversal of a global pattern, almost every Arab country today is less free than it was 30 years ago. There are few countries in the world of which one can say that."

To address the frequent claim that economics of poverty is the driving force of such changes, or the opposite that oil wealth might be the driving force, he clearly articulates the absence of economic motivation through the statement, "If poverty … [was responsible] in most of Arabia, wealth … [was responsible] in the rest of it.… All that the rise of oil prices has done over three decades is to produce a new class of rich.…"

What is the reason for this dramatic difference? The causes are clearly not just economic; they are primarily cultural, with religion as the driving force. Among the key elements of Islamic culture that are relevant to this trend is the accepted understanding that the state is responsible for imposition of cultural norms within an Islamic society. This is directly counter to the promotion of individual freedom and diversity that is characteristic of Western thought and is at the center of systems that are not centrally controlled. This difference also leads to a local incompatibility of the socio-cultural systems.

This incompatibility of local social perspective can be understood as analogous to the incompatibility of oil and water. When the two are mixed they tend to separate. As they separate, larger regions of one and the other form and the ongoing process of separation occurs at the boundary between the two. A process of pattern formation takes place, similar to the discussion of fads in Part I of this book. The boundaries become better defined, smoother and flatter over time. When we think of this process, the analogy to ethnic violence as it has occurred in many parts of the world appears clear. Indeed, we can consult lists of the locations where ethnic violence is currently occurring or has been taking place over recent years and we find that a large majority of them are located along the boundary between Islamic areas and other areas. It is important to emphasize that which side is the aggressor is not the issue in this context. It is also not a question of determining which side is in the right or wrong. The key is recognizing the underlying process that is taking place. In order to do so we must see the connection between all of these conflicts rather than considering any one of them in particular. Each one has a specific and detailed history with local historical aspects that are not shared with other conflicts.

Violence at the boundary between Islam and Christianity (Western and Orthodox) occurs in Bosnia, Chechnya (part of Russia), Philippines, and Indonesia. It occurs between Islam and Hinduism in Kashmir (part of India). Violence in Africa includes conflict between Islam and various local cultural groups that are becoming increasingly Christian. Violence between Islam and Judaism occurs in Israel.

The recognition of the importance of the boundary between Islamic and non-Islamic areas resonates with but is different from the ideas of Sam Huntington. His book, *The Clash of Civilizations and the Remaking of World Order*, describes the relevance of conflict between the major different cultural regions of the world just at the time when the conflict with the Soviet Union had ended. The next conflict, Huntington argued, would be between the different "civilizations" of the world. While he considered the conflict between civilizations generally, he emphasized the conflict between Islam and others. Here, this conflict is reconsidered. Conflict is not intrinsic to the relationships between the civilizations, but rather results from a need to differentiate between local conditions in the different cultures and thus establish clear boundaries between them.

The key to understanding the incompatibility of Islam and other cultural systems lies in understanding the characteristics of organization and the level of uniformity. Other systems have a greater respect for individual differences and diversity. Islam insists on a significant level of conformity to cultural behavior patterns. Such conformity must be imposed collectively, leading to the need for Islam-based institutions, including desire for an Islamic state.

Some may argue that what is needed are educational efforts to moderate religious views. However, this approach reflects an intolerance for both individual and group level choices. Tolerance at the individual level is also not the same as tolerance at a societal level. Should a society have the right to impose uniformity? Because Western culture values freedom of choice at the individual level, it does not tolerate the larger scale choice of the Islamic culture. We see here directly the conflict between larger scale and finer scale behavior discussed in the first part of this book. The cultures are intrinsically incompatible because of the primary scale at which freedom of action is allowed.

The implication of this analysis is that separation of these two cultural systems is likely to continue. If separation continues, then, as the boundaries between the two systems become clearer, the problems of ethnic violence will diminish. Indeed, the best way to inhibit ethnic violence is to promote the separation rather than discourage it. A key question then becomes how to structure the boundary between the systems. For example, what level of commerce and interactions will be possible? The answer is likely to differ in different parts of the world. In general, however, many forms of trade of commercial goods should be possible.

The ideal that everybody should be able to live together in harmony has here a different form of realization than that at the individual level. The vision presented here is the harmony of cultures existing together at a larger scale of organization—not of individuals mixing and interacting freely throughout the world. Two cultures can coexist peacefully when they have the appropriate interactions and the appropriate separations.

Terrorism and global military actions

The local interactions of ethnic violence at the boundary of Islamic and non-Islamic regions also have a global (not local) aspect: terrorism and the asymmetric War on Terrorism and global military actions. Global terrorism manifested itself in 9/11 and in earlier bombings (hijackings, etc.) aimed at Western entities. These are asymmetrically countered by police actions around the world, limitations on travel and financial flows, and military actions in Afghanistan and in Iraq.

As discussed in Chapter 9, the War on Terrorism is a highly complex one requiring diverse actions in many places around the world. Among the actions that are needed is a reduction in the occurrence and severity of local ethnic conflicts. These local wars create regions of lawlessness and violence that breed terrorists, motivate the formation of organized terrorist groups, and provide bases for operations. Groups formed in regions of conflict support each other in developing international activities including training and coordination of terrorist actions.

Terrorism is also linked to the process of separation between Islamic and non-Islamic populations in other ways. This includes practical as well as intentional aspects of the terrorist actions. Practically speaking, terrorism increases the difficulty and risk of travel for non-Muslims to go to Islamic countries, and for Muslims to go to non-Islamic countries. Also, one of the stated demands of terrorists is the departure of non-Muslims from Islamic countries. Indeed, a key stated reason for the terror against the U.S. is the departure of military personnel from the Islamic holy land of Saudi Arabia. The strong sense of a need for total separation is also clear from the reception of U.S. forces in Iraq even by those who have been freed from the oppressive regime of Saddam Hussein. These effects manifest the underlying forces toward separation.

The existence of such a widespread desire for separation may also undermine many aspects of the current strategy (or other strategies that might be adopted) in the War on Terrorism. Actions that promote more individuals to adopt a course of violence will be counter productive. Acting in a way that respects the underlying social concerns but still opposes terrorist activities, would be much more effective. Such actions will avoid increasing terrorist recruitment and formation of new terror organizations. In particular, strategies that involve placing non-Islamic individuals into Islamic countries should be considered a last resort.

The current conflict in Iraq can also be considered in this context. There are many and varied political approaches to this context. For some, this war is an extension of the 1991 Gulf War to expel Iraqi military forces from Kuwait. The rapid and successful completion of the objectives and the positive reception to the U.S. involvement in 1991 led many to have expectations for similar outcomes today. It is important to develop a better understanding of the key differences between the current Iraq war and the 1991 Gulf War.

To analyze some of these differences, we can focus on the connections between people. In the Gulf War the enemy of the U.S. was an occupying Iraqi army located in Kuwait. In the Iraq war the proclamations about the war against Saddam Hussein and his dictatorial and ruthless regime were couched in the same way. The idea that the U.S. would serve as liberators of the Iraqi people from an oppressive regime seems very reasonable.

Yet, the military force that opposed the U.S. in this conflict had fathers and mothers, siblings and children—members of the population that were being "freed" from them. Those opposing military forces were much more connected to the people of Iraq than are the U.S. forces. No matter how violent a regime was present there, this factor implies that there are many individuals who will feel that the U.S. is not a liberator but an alien entity. When this is combined with the deep internal divisions within Iraq (between oppressed and oppressor, religious and secular), and the severe cultural clash between all of these groups and the U.S. forces, it is easy to recognize that the situation is not easy to control.

More significantly, when we consider the historical role of Saddam Hussein, we notice that in the past the U.S. was his supporter. Why would this be the case? The reason is that Saddam was opposing the extreme religious government of Iran. Today there is a sense that Saddam developed chemical and biological weapons for war against the U.S. Without justifying such weapons, we should recognize that this is not the case. Saddam developed these weapons in his battle with Iran, a brutal regime. Internally in his country, Saddam was suppressing the same fundamentalist Islamic groups that made Iran the country it is today. Thus, Saddam's historical role has been as a secular military dictator in opposition to fundamentalist Islamic forces and this is a pattern we find repeated in other parts of the Islamic world. The brutality of Saddam's regime is well documented. Still, a classic analysis of friends and enemies would place him against, not with, the most virulently anti-Western groups. Now, the U.S. has "rescued" these anti-Western groups by invading Iraq. While some may think they would be grateful, given their fundamental views on the world, we should not be surprised that they have limited interest in welcoming the U.S. Moreover, Iraq's opposition to Iran has been diminished, providing opportunity for Iran to focus on its opposition to the West. Of course, in the context of the cultural divide, both the secular and the religious Islamic groups may be anti-Western. What we should realize, however, is that a natural course of events that may follow from the ouster of Saddam would lead to another government like Iran's where religion plays the role of suppressing individual freedoms. Alternatively, and somewhat less likely, is the development of another kind of dictatorship. Democratization of Iraq, that some would like to believe possible, is not likely in the context of these forces.

This, however, is not even the greatest problem. The greatest immediate problem is the ongoing intimate engagement between U.S. forces and Islam in Iraq. This contact is directly counter to the need for separation, and a great source of irritation, like a mixing of oil and water. The most natural outcome of such an engagement is the development of a new area of disorder that serves as a substrate for terrorist activities. This is the greatest source of concern when the larger pattern of separation is considered!

Conclusion

Ethnic violence and the related terrorism are not necessarily rooted in conventional military conflict. It is a cultural/political/social challenge. While many people may view these conflicts in terms of desires for conquest, the underlying pattern can be viewed as one of global pattern formation and differentiation. It seems reasonable, therefore, to see the conflict as a need for separation. In the meantime, the U.S. is fighting this separation and appears to be following an underlying assumption that individuals (Western or not) should have the freedom to be anywhere. Ultimately, it is this priority that seems to be a losing ideological battle.

The existing national boundaries generally do not align with the cultural boundaries that are forming. In order to avoid violence we must promote the separation of groups that are currently mixed or are subject to common governing structures. This may involve negotiating new administrative regions with clear boundaries (geographic or behavioral), possibly even physical barriers or guarded borders. In many cases in order to provide a clear separation it may be necessary to provide financial help or incentives for individuals to move, or even to negotiate the movement of larger groups of people. Each circumstance should be considered in its own historical and cultural framework, but with attention to the global patterning process underway. The expectation that distinct approaches to ways of life will be able to reside side by side is not unreasonable as long as the contact between them is bounded in its scope. Commerce and trade can occur across cultural boundaries and respect the ideological divides. Diversity of cultures living together peacefully is not the same as having all individuals peacefully mixed together. However, it is a reasonable view of the ultimate nature of global peace.

Chapter 17

Conclusion

To solve complex problems we must create effective complex organizations. The underlying challenge of this book is the question: How do we create organizations that are capable of being more complex than a single individual? Living with complexity is challenging, but we can and should clearly understand the nature of how it can be done, both for individuals and organizations. The complexity of each individual or organization must match the complexity of the task each is to perform. When we think about a highly complex problem, we are generally thinking about tasks that are more complex than a single individual can understand. Otherwise, complexity is not the main issue in solving it. If a problem is more complex than a single individual, the only way to solve it is to have a group of people—organized appropriately—solve it together. When an organization is highly complex it can only function by making sure that each individual does not have to face the complexity of the task of the organization as a whole. Otherwise failure will occur most of the time. This statement follows quite logically from the recognition of complexity in problems we are facing.

Our experience with organizing people is for large-scale problems that are not very complex. In this case the need for many people arises because many individuals must do the same thing to achieve a large impact. In this old reason for organizing people, a hierarchy works because it is designed to amplify what a single person knows and wants to do. However, hierarchies (and many modifications of them) cannot perform complex tasks or solve complex problems. Breaking up (subdividing) a complex task is not like breaking up a large scale task.

The challenge of solving complex problems thus requires us to understand how to organize people for collective and complex behavior. First, however, we have to give up the idea of centralizing, controlling, coordinating and planning in a conventional way. Such efforts are the first response of almost everybody today because of the effectiveness of this approach in the past. Instead, we need to be able to characterize the problem in order to identify the structure of the organization that can solve it, and then allow the processes of that organization to act. The internal processes of that organization can use the best of our planning and analysis tools. Still, ultimately, we must allow experimentation and evolutionary processes to guide us. By establishing a rapid learning process that affects individuals, teams and organizations, we can extend the reach of organizations, allowing them to solve highly complex problems.

I appreciate that I am only one human being and my understanding of the world is consequently quite bounded. Still, it is reasonable to hope that some of the concepts discussed here may be of use to you. Others will complement or contradict me as necessary.

The basic concepts that I hope to have contributed an appreciation for are as follows:

- The functional importance of independence, separation and boundaries as counterpoints to the importance of interdependence, communication and integration;
- The trade-offs in scale and complexity, where increasing the set of behaviors possible at one scale (complexity at that scale) requires a reduction in complexity at other scales;
- The need for matching the complexity of the system at each scale to the complexity of the environment (task) at the same scale for the system to be successful;
- The diverse nature of distributed networked systems that are not all the same thing (contrast, for example, the immune system and the nervous system), but can be understood from the same general principles;
- The essential complementarity of competition and cooperation at different levels of organization;
- The constructive nature of both competition and cooperation in forming complex systems;

- The limitations of conventional planning in creating and managing complex systems and the essential importance of planned environments for evolutionary processes;
- The practical utility of fundamental complex systems ideas;

Slightly less apparent but no less important are the recognition and appreciation of:

- the profound paradoxical importance of individual and group differences as a universal property of complex systems;
- the significance of specialization in effective collective behavior, including specialization of individuals and specialization of large subsystems;
- the remarkable emergent behaviors that combine simple capabilities to allow dramatic system capabilities;
- the universal nature of patterns of collective behavior, which serve as elementary building blocks of complex systems just as atoms do;
- the ubiquity of pattern forming processes, differentiation, and particularly local-activation long-range inhibition mechanisms for such patterns.

Finally, along with the recognition of complex problems that we continue to face in this world, we have also pointed out the increasing complexity of society. This increasing complexity implies great capabilities. Indeed, it suggests that we, together, are becoming remarkably effective at solving complex problems in a complex world.

The following sections briefly review the recommendations that we have made about diagnosing and solving several complex problems facing the world today. In addition, we conclude the book by providing examples of successful systems, demonstrating the power and utility of adapting a complex systems approach.

Diagnosing systems

When we are faced with what seem to be intractable problems today, diagnosing the problem and identifying why it exists is a first step toward solving it. From experience, the way we have been trying to solve most problems contributes greatly to their existence and difficulty. Unfortunately, we are still responding to societal problems by centralizing authority and imposing the will of one person. This is the standard way we try to solve complex problems and the reason these efforts don't work. We also use outdated metaphors when discussing these problems. As we discussed in the section on military conflict, we often use terminologies like the "War on Poverty" and the "War on Drugs" to describe these efforts (Soon there will be a "War on the Education System"). Ironically, even military conflict today often does not follow the traditional concept of war. The U.S. military has demonstrated in Afghanistan that it has an understanding of complex warfare. This is not just because people in the military have studied the concepts of complex systems (which they have), but because they have a much shorter feedback loop for learning. If the military is doing something that doesn't work, they tend to learn about it much faster than others do. Still, the war in Iraq demonstrates that the lesson of complex warfare is not universally understood. It is important to emphasize that political leaders of military forces may not have the benefit of military experience and thus direct the military to actions that are counter to this experience.

Learning from the experiences that we have gained by using the traditional approach to solving complex problems has been much of the subject of this book. We identified the failure of the standard approach in health care, education, engineering, third world development, and the War on Terrorism. The symptoms of failure in these cases have to do with a sense of crisis arising from widespread local problems. In each of the examples we have discussed, the problem is everywhere, but local conditions prevail, confounding the conventional strategy of centralization. This is complexity in action.

Indeed, the first step in solving a complex problem is developing an understanding of the complexity profile of the system: the way that complexity and scale exist in the tasks that need to be done. This may be summarized by identifying the complexity at each scale. The complexity profile captures the degree to which actions of the system are (or need to be) repetitive, and to what degree they need to act in response to local conditions at different places, or over time to different instances.

Quite generally, an analytic approach should focus on distinguishing those processes that are large scale and therefore can be performed efficiently, from those processes that are highly complex, requiring individual specialization or even teams to perform. When large scale tasks are identified, then one can adopt the traditional approach of centralizing authority, instituting standards, imposing uniformity, planning upgrades and improving efficiency. When complex tasks are identified, then one should adopt the complex systems evolutionary approach of distributing decision, action and authority, setting functional goals and directions for improvement, supporting individual initiative, measuring effectiveness in the field, instituting redundancy, forming cooperative teams, and creating rules that promote competition with performance feedback at the functional team level. Appropriately, it is more difficult to address complex problems as there is no one universal organizational structure that will work for all cases. However, our discussion has identified strategies for determining which organizational forms can work by analyzing information flows. Moreover, we can allow them to form without analysis using an evolutionary process.

The medical system and the education system are both characterized by high complexity fine scale tasks that are demanding on the individuals (doctors, teachers) who perform them. Engineering projects of real time response systems, and international efforts to develop functional societies, are both engaged in the desire to create remarkably complex systems. The War on Terrorism appears to be concerned with a locally hidden terrorist network, however, underlying its existence appears to be a global dynamic of billions of people in a collective socio-cultural process. In the following sections we summarize briefly these problems.

Health care and education

For the health care and education systems, we are engaged in the improvement of systems that are internal to the U.S. (or other) society and have specific functional roles. Our primary concerns are effectiveness and efficiency of these systems. Today, both systems are changing toward a more uniform approach. The starting points, however, are quite different, with health care much more individualized than education. In both cases a complex systems analysis suggests that most of the tasks are highly complex and a uniform approach will not work.

The reason for the existing difference in the health care and education systems can be understood from the current way we learn whether tasks are being performed effectively. The medical system focuses on tasks whose success is readily observable, often life or death, while the education system focuses on tasks that take a long time for us to see their effect. The result is that we have a good way to evaluate rather quickly whether an individual doctor is doing a good job, but poor ability to evaluate whether an individual teacher is doing a good job.

The consequence is that the medical system has a more highly complex system based upon highly specialized physicians who can perform highly complex tasks. It has a routing system to direct patients to the right doctors.

The education system also has some specialization, but it appears gradually, mostly in colleges and graduate school. There is little specialization in elementary and high school, though a small number of magnet schools do form some exceptions. In most middle and high schools there is limited specialization of teachers, and much more limited routing for students.

Moreover, the medical system has a highly rigorous training process to produce physicians who are capable of their highly complex tasks. In comparison, the teacher training system is much less demanding and extensive. There is substantial selection by the recipients of care of their doctors, but very little selection by children/parents of their teachers. (In recent years there is a tendency for health care plans to reduce physician selection in the medical system. Like many other changes that have been made, this is a trend with problematic implications.)

In terms of total dollars spent, the rapidly growing U.S. health care budget recently passed 15% of GDP. Spending on education has been relatively constant at 6-8% of GDP. The medical system is oriented toward high cost exception handling of individual cases, while the education system is oriented toward a low-cost continuous and uniform process.

The medical system is being driven toward greater uniformity, through regulation and control over the behavior of individual physicians, primarily in order to improve efficiency. This, however, leads to a loss of effectiveness in the (increasingly) high complexity tasks being performed. The low quality and high error rates are further motivating the efforts to regulate and control the behavior of physicians. Meanwhile the education system is being driven toward greater uniformity primarily in order to improve quality. These are classic examples of going in the wrong direction because of a misunderstanding of the origins of the problem.

In both cases a complex systems approach suggests that we should use local competition to improve quality of complex tasks, and use uniformity for efficiency in large scale tasks.

While the health care system already has various forms of competition that improve care by individual physicians, these should be augmented by improved feedback and emphasis on the role of teams for tasks that are too complex for traditional physician specialization. At the same time, identifying those aspects of health care that are large scale will give significant opportunities for achieving higher efficiency through population care, rather than individualized care. This is particularly important because of the need to relieve the financial pressures on the performance of complex tasks, and the importance of prevention that can be achieved at least in part as a large-scale task.

The education system requires more radical changes to achieve the necessary increase in individualization. The main change needed is to develop a higher complexity local specialization that will allow effectiveness in the complex task of education. The key is to have both an increase in specialization in teaching, and a process for routing of children to the educational environments in which they will flourish. The attention to individual differences through developing individual skills will enable children to be effective in increasingly diverse professions. Measures must be developed that can allow for diverse and highly demanding criteria of success.

Both of these systems need to follow the general principle of properly matching their organizational structures to their tasks. The medical system should develop an additional structure that is capable of addressing large-scale needs of the population (preventative care and screening tests). The education system, on the other hand, should develop a highly complex system capable of effectively educating each child. Furthermore, both systems should develop local competitive evolutionary processes, which improve the quality of the system performing complex tasks.

Engineering and international development

The existing approaches to the engineering of highly complex systems for the government, civilian and military, as well as for large corporations, are based upon a directed planning approach. Planning is also the approach being used for interventions that are designed to promote development of functioning economies and societies in the third world. The great frustrations that have been experienced with failed projects in both spheres are ample manifestation of the inadequacy of planning for addressing the creation of effective complex systems.

To overcome these problems, we recommend again the approach that was recommended for improvement of the health care and education systems. First one should consider the scale and complexity of the tasks involved. This analysis should reveal the importance of system structure at different scales and how they match the tasks that must be performed. Second, resort to conventional planning design and implementation for not overly complex tasks. Third, set up a context for and rely upon evolutionary processes to create highly complex aspects of the systems that are desired.

In both the engineering and development contexts, the devotion to planning arises from an expectation that careful planning guarantees success. Specifically, it guarantees that we will get what we want, when we want it, and for a price we can determine in advance. Complex tasks and the complex systems that are necessary to perform them do not allow for such certainty.

What can we use to substitute for the feeling of security? An evolutionary process of accumulating parallel incremental changes can offer an ability to invest much smaller sums of money up front than the full project requires, with significantly more rapid feedback on improvements that are being achieved. Any project that is created through an evolutionary process must be designed to provide improvements that are noticeable and important to the functioning of the system in a comparatively short time. This is in sharp contrast to the conventional planning based process where one must wait a long period of time before any information is received about the utility of a new system. In a circular fashion, this delay is itself a reason that planning is considered crucial to avoid large investments without knowing how the effort is progressing. Evolutionary improvements that occur early and often, can provide confidence that ongoing investments will be fruitful.

Military conflict, terrorism and ethnic violence

We started the book by discussing military conflict because it is easy to visualize the key insights of complex systems in this context. In particular, how large scale forces composed of tank divisions were effective in the Gulf War, but high complexity Special Forces were effective in Afghanistan.

Today the world is experiencing a variety of military conflicts, the war in Iraq, War on Terrorism, and many less widely reported conflicts typically involving ethnic violence. The U.S. approach to the war in Iraq is a manifest mismatch of scale and complexity, with complexity residing in the hands of those diverse individuals and groups who oppose U.S. forces. If the U.S. military is to fully benefit from its extensive experience in complex conflict, there is a need to formalize this knowledge, so that in the future wishful thinking does not trump knowledge.

The War on Terrorism includes terrorist forces that have both high complexity and a wide diversity of available targets. This is what gives them an opportunity for causing substantial damage. Opposing them, however, is also a highly complex array of organizations including: military, intelligence, law enforcement, diplomatic and other organizations from many countries and localities. While some need surely exists to develop joint activities, perhaps the main danger is from efforts to centralize and coordinate these forces in a conventional way.

In this book, however, we have proposed a broader view on the global conflicts that exist today in the hope of promoting a solution that can work. The key to this view is a recognition of a very large scale phenomenon that comprises the collective action of billions of people on earth. This is the development of socio-cultural domains, whose members do not readily mix. This is particularly true for the ongoing socio-cultural separation of Islamic populations from non-Islamic populations. This separation has been manifest in the ethnic violence at the boundaries between these two population groups. Our objective is not to assign blame for these conflicts, as any local conflict can have diverse origins. Instead, the objective is to suggest that a proactive effort can be made to prevent further conflict. A necessary step towards reducing ethnic violence is establishing clear boundaries between the groups, so that socio-cultural differences can coexist without friction. Current social boundaries established by historical processes do not follow the need for boundaries as they exist today. Redrawing the boundaries of control and interaction within countries or changing country borders would be an important step. Seeking means for voluntary separation of minorities where violence is likely would be much better than forced separation or the pattern of ethnic violence as it currently exists and is likely to continue. Early action to separate groups will preempt ethnic conflict and much loss of life. The use of boundaries to separate socio-cultural systems does not mean that relevant economic and business contacts cannot exist and be successful.

Here as in other topics we have addressed, the importance of taking a multiscale and multilevel view is manifest. Individual freedoms always exist at the expense of collective behaviors, collective behaviors always exist at the expense of individual freedoms, and there are trade-offs between any two levels of organization. The respect for individual differences that is emphasized in the U.S. is not sufficient. It is necessary to develop an appreciation of differences in social systems at all levels of organization.

When systems work

In this section we will review several examples that illustrate how planned evolutionary competitive/cooperative environments exist and are remarkably successful in the economy today, demonstrating what can be achieved through a complex systems framework to solving problems.

Marshall Plan

The Marshall Plan, which followed World War II, was the basis of the rapid recovery of western Europe (including both sides of the conflict, i.e. West Germany and Italy, as well as Great Britain, France, Netherlands, and other countries) after the devastation of the war.[91] Its success stands in marked contrast to the policies that followed World War I, which were imposed by the victor on the defeated, were primarily punitive in nature, and whose devastating economic consequences are often blamed for the growth of fascism leading shortly thereafter to World War II.

The Marshall Plan was intentionally announced through what was an unusual method of presentation—a short lecture at Harvard University—and can be summarized as 'there is a clear need, let us know how we can help and we will.' This approach explicitly rejects planning and central control in favor of actions motivated by the local understanding of those who are most directly involved. Indeed, many and diverse forms of assistance resulted. The Economic Cooperation Administration that implemented the Marshall Plan provided direct financial support, loans and loan guarantees for a wide variety of large and small projects[92] and is justifiably credited with enabling the rapid recovery and subsequent growth of the European economy. Further, avoiding the creation of an extended dependency on assistance, the Marshall Plan was limited to four years.

The deep understanding of the reasons for such a policy are manifest in Marshall's speech itself. These include a recognition of the importance of the internal structure of economic interactions and relationships, the interdependence of economic and socio-political instability, the inadvisability of imposing an external solution, and the remarkable complexity of the world. The following excerpts from Marshall's lecture demonstrate this understanding:[93]

> I need not tell you gentlemen that the world situation is very serious.... [O]ne difficulty is that the problem is one of such enormous complexity that the very mass of facts presented ... make it exceedingly difficult ... to reach a clear appraisement of the situation.... [T]he physical loss of life, the visible destruction ... was correctly estimated, but it has become obvious during recent months that this visible destruction was probably less serious than the dislocation of the entire fabric of European economy.... Long-standing commercial ties, private institutions, banks, insurance companies and shipping companies disappeared.... The breakdown of the business structure of Europe during the war was complete.... Europe's

> requirements for the next three or four years of foreign food and other essential products—principally from America—are so much greater than her present ability to pay that she must have substantial additional help, or face economic, social and political deterioration of a very grave character.... It is logical that the United States should do whatever it is able to do to assist in the return of normal economic health in the world, without which there can be no political stability and no assured peace. Our policy is directed not against any country or doctrine but against hunger, poverty, desperation and chaos. Its purpose should be the revival of a working economy in the world.... Any assistance that this Government may render in the future should provide a cure rather than a mere palliative. Any government that is willing to assist in the task of recovery will find full cooperation, I am sure, on the part of the United States Government.... It would be neither fitting nor efficacious for this Government to undertake to draw up unilaterally a program designed to place Europe on its feet economically. This is the business of the Europeans. The initiative, I think, must come from Europe. The role of this country should consist of friendly aid in the drafting of a European program and of later support of such a program so far as it may be practical for us to do so....

Intentional markets

The ability of "free markets" to enable the exchange of goods without central coordination has been the basis of development of much that we appreciate in the world economy. However, markets often occur in a framework that has some amount of central planning and coordination. These systems might be called 'intentional markets' and they correspond to our discussion of complex systems approaches to developing frameworks in which evolutionary competition and cooperation can provide effective systems. The New York Stock Exchange serves as an example of such a system.

The existence of the New York Stock Exchange (NYSE)[94] is a central feature of the economic activity in the U.S. and serves companies and investors throughout the world. Its roots trace to the Buttonwood Agreement in 1792, which was signed by 24 stockbrokers, and subsequently the formal organization of New York Stock & Exchange Board in 1817 with a constitution that dictated the rules of exchange. In effect, these rules establish a collaborative framework for competition.

The exchange provides a list of companies, numbering 2,750 at the end of 2003 with a total value of 17.3 trillion dollars. Shares in these companies may be traded by members of the exchange who represent investors, i.e. investors can pay commissions to members of the stock exchange to buy and sell stocks for them. Every transaction is a competition between those who want to sell shares, and at the same time a competition between those who want to buy shares.

Someone who owns shares in a company has the right to sell them, or to buy shares from another owner. Today, investors generally choose to have their money and shares held for them in an investment account at a company that either is a member of NYSE or that arranges trades through one of the members. The key reason that the NYSE is an effective system is that the cost of performing a transaction is low, so the commissions are low and many people choose to buy and sell in this market. The total value of trades was almost 10 trillion dollars in 2003.

One of the key features of this market, like other markets, is that it can deal effectively with individual needs. In particular, it can perform very large and very small transactions. For example, the largest single block of stocks sold in the last few years had a value of over 5.4 billion dollars. On the other hand, it is reasonably common to find trade commissions that are about $10, and there are no lower limits to the amounts that can be traded. Members of the exchange continue to compete with each other in a wide range of services. For example, they were quick to adopt Internet based services when the Internet became commonly used in the last ten years.

Intentional markets illustrate how setting up a structured cooperative framework for competition enables complex tasks to be accomplished. Such a market is a collaboration at the market level, a competition between sellers and buyers, and fosters remarkably effective cooperation within the organizations that are acting as sellers and buyers. The competitors become highly capable at their tasks and the market as a whole works remarkably well for society.

VISA International/MasterCard International

Most people do not know that the largest corporation in the world, measured by revenue, is VISA international. VISA is not a publicly traded corporation. It is an organization that is owned by its members, the companies that issue credit cards that carry the VISA name. There are about 21,000 members. Many members are banks but this is not exclusively the case. Sears, Disney and other companies are also members. MasterCard

International is a similar member-owned corporation, with about 25,000 members. Their combined revenue, given by the total transactions on credit cards is approximately 4.3 trillion dollars, 3 trillion by VISA and 1.3 trillion by MasterCard. What is particularly remarkable is that they account for approximately 13% of the total of personal purchases worldwide. Personal purchases comprise about half of the global economy. The other 87% of personal purchases is mostly cash and checks. The proportion of purchases through these credit card organizations has grown steadily and is expected to continue to grow into the future.

MasterCard International was formed as a bank cooperative in 1966 (Interbank Card Association), and changed its name to MasterCard International in response to the formation of VISA International in 1976. VISA International was created in 1976 through the adoption of an agreement by member companies. It superseded a system controlled by Bank of America. The founder, Dee Hock, is aware of the connection between complex systems and his efforts in creating the cooperative. His recent autobiographical book is titled *Birth of the Chaordic Age*, where Chaordic is a word that combines "chaos" and "order."[95]

Both VISA and MasterCard were created as a framework in which members can compete with each other while cooperating in setting the framework for the competition. This is similar to an intentional market, and is a natural realization of our discussion of an evolutionary environment. Not all of the features of this system may be exactly what theory would suggest, however, the correspondence is strong.

While the size of these organizations as measured by total transactions is impressive, what is particularly important for our understanding of their success is their ability to perform a service function in society that is pervasive, global, and has many local aspects that are important. Indeed, the pervasive nature of VISA and MasterCard has made their organizations almost unnoticed, like wallpaper, part of the environment in which we live.

Open source movement

The open source movement is a collection of people who are engaged in developing computer software. It is called the "open source" movement because everyone can have access to the original programs, often called "source code," from which applications are made. The source code is free, typically by download from the Web. To participate, a person must respect an agreement that forms a framework for the activities of the community.

Its primary feature is that it imposes general access to modifications made in the programs.

Thus, this is a self-defining community of computer programmers who can all work on improving programs by rewriting their source code. Through a process of approval that is somewhat centrally directed, in the sense that respected authorities have arisen within the community, improvements are designated for inclusion in widely used distributions of the software. Multiple versions, however, can and do exist. Therefore, there are multiple levels of competition, between individuals who want to have their innovations included in the software distributions and between versions of the software. The competition is not motivated by financial gain, but perhaps by the reward of knowing (or having public recognition) that one's contribution is being included and used by many other people. Financial gain is not excluded in the context of community activity. Anybody can sell products based on these programs and provide services like consulting and support for other users.

There are people who are strictly users and, in that sense, they are consumers of open source products. However, many of the users are also the producers of the programs, so the distinction is not generally clear.

Over the past few years, one of the systems that the open source movement has been developing, called Linux, has become widely used in computer servers that underlie the Internet, challenging both Sun Microsystems, and Microsoft, more conventional corporations that have devoted substantial efforts to creating software. In 2002, IBM reported[96] that it invested one billion dollars to make their computers compatible with Linux, and that it had more than recovered the investment in sales and services provided to users. In 2003 the total of sales of servers running Linux has been reported to be 2.8 billion dollars, 6% of the server market.[97] Both sales and market share are growing rapidly with the first quarter of this year having one billion dollars in sales, which is 8% of the overall market. Moreover, since Linux servers tend to be lower in cost, this corresponds to almost 15% by number of servers sold.

This is a remarkable example of how this community has together developed a product that is challenging proprietary corporate software in many markets. Why should the open source movement be challenging these companies, particularly Microsoft which is by far the largest software company and has a history of success against other corporate competitors? The effectiveness of evolutionary processes provides an understanding. The key is not just where each of the competitors lies today, but that the open source movement is advancing faster than Microsoft can advance.

The open source community is an example of the evolutionary process we have been describing. The rules of the community establish a framework for interaction and competition, but do not specify or plan the software that will result. The rate at which many nearly independent individuals introduce innovations into open software is very fast. Most of them are likely to be rejected immediately or after a short time (as is also the case for the introduction of new products into the food market). The ones that are selected for eventual inclusion are likely to be among the best. The effectiveness of the community demonstrates the advantage of an evolutionary approach over planning.

New developments in software in a wide range of areas continue to occur. What is particularly remarkable is that the open source movement is beginning to gain ground on Microsoft in its strongholds where its monopoly power should prevent others from gaining any ground at all.

Summary

In this book we have described principles that clarify how complex systems can cope effectively with complex environments. The remarkable complexity of our society often overwhelms us but has the potential of creating an increasingly protective and productive environment for each individual, linked to an increasingly effective collective behavior.

The difficulties we face in providing essential aspects of well being, education, health care, engineering and economic development, are happening largely because we don't recognize the power of our complex collective. As we gain this insight, these difficulties will surely be resolved.

Notes and References

PART I CONCEPTS

Chapter 1 Parts, Wholes, and Relationships

1. Looking at the parts of a system is like knowing that all books in English are formed of 26 letters, capitals and punctuation.
2. The following textbook will be referred to frequently in this book, henceforth it is called DCS:
   - Y. Bar-Yam, *Dynamics of Complex Systems* (Perseus Press, 1997). http://www.necsi.org/publications/dcs
3. For a discussion of various forms of emergence see:
   - Y. Bar-Yam, A mathematical theory of strong emergence using multiscale variety, Complexity **9:6**, 15–24 (2004).
4. DCS, pp. 91–95.

Chapter 2 Patterns

5. The mathematical description of pattern forming systems was initiated by Alan Turing:
   - A. M. Turing, The Chemical Basis of Morphogenesis, Philosophical Transactions of the Royal Society B (London) **237**, 37–72 (1952).

   The simple pattern-forming rules that are described in the text using spatial arrays of elements are called Cellular Automata and were introduced by John von Neumann:

- J. von Neumann, *Theory of Self-Reproduction Automata*, edited and completed by A. Burks (University of Illinois Press, 1966).

Such rules were further developed in the 1970s. Many people are familiar with a particular rule called Conway's "Game of Life." Original scientific articles by many authors, and an extensive bibliography, are collected in the book:

- *Theory and Applications of Cellular Automata*, edited by S. Wolfram (World Scientific, 1983).

A pedagogical introduction to Cellular Automata can be found in DCS, pp. 112–145; and a discussion of pattern formation in DCS, pp. 621–698.

6. A better picture of what is happening in mammals includes the recognition that cells that produce pigment are themselves mobile. As they move, the cells have a tendency to aggregate and/or repel other such cells. The attraction and repulsion leads to the patterns that are formed. The precise mechanism may vary, but the general principles associated with a tendency to similar behavior nearby, and different behavior farther away, characterizes the formation of such patterns. See DCS, pp. 621–698.

Trademarks referenced in this chapter:

- Pokémon is a registered trademark of the Nintendo Corporation.
- Beanie Babies is a registered trademark of Ty, Inc.

Chapter 3 Networks and Collective Memory

7. A collection of articles describing the origins of mathematical models of neural systems can be found in:

- *Neurocomputing*, edited by A. Anderson and E. Rosenfeld (MIT Press, 1988).

The original Hebbian model of imprinting is found in:

- D. O. Hebb, *The Organization of Behavior* (McGraw-Hill, 1949).

A simple attractor/associative network model was introduced in:

- J. J. Hopfield, Neural networks and physical systems with emergent collective computational properties, Proceedings of the National Academy of Sciences (USA) **79**, 2554–2588 (1982).

A pedagogical discussion of attractor and feedforward networks can be found in DCS, pp. 295–328.

8. The problem of understanding creativity is discussed in the following collections:
   - *The Creativity Question*, edited by A. Rothenberg and C. R. Hausman (Duke Univ. Press, 1976).
   - *Creative Thought*, edited by T. B. Ward, S. M. Smith and J. Vaid (American Psychological Association, 1997).

   The subdivided architecture of the brain is described in:
   - M. S. Gazzaniga, R. B. Ivry and G. R. Magnum, *Cognitive Neuroscience*, 2nd edition (W.W. Norton & Company, 2002).

   The universal importance of subdivision in complex systems is described in:
   - H. Simon, *Sciences of the Artificial*, 3rd edition (MIT Press, 1996), Chapter 8.

   The relationship between subdivision in the brain and creativity is discussed in:
   - DCS, pp. 328–419.
   - Y. Bar-Yam, Why (partially) subdivide the brain, NECSI Research Report YB-0008 (1993).
   - R. Sadr-Lahijany and Y. Bar-Yam, Substructure in Complex Systems and Partially Subdivided Neural Networks I: Stability of Composite Patterns, InterJournal of Complex Systems [1] (1995).

   The idea that a hard-wired language acquisition system must exist in the brain to account for the universality of grammar and the observation that children are not exposed to enough language to explain the learning of language was proposed in:
   - N. Chomsky, *Aspects of a theory of syntax* (MIT Press, 1965).

   The role of subdivision in the brain in language acquisition is described in the references above (DCS, pp. 328–419, and Y. Bar-Yam (1993)).

Chapter 4 Possibilities

9. The mathematical theory of communication and information was described by:
   - C. E. Shannon, A Mathematical Theory of Communication, Bell Systems Technical Journal, July and October 1948; reprinted in C. E. Shannon and W. Weaver, *The Mathematical Theory of Communication* (University of Illinois Press, 1963).

   Various studies of the concept of complexity have used the concept of information developed by Shannon or a different approach developed independently by Salomonov, Kolmogorov and Chaitin:

- R. J. Solomonoff, A Formal Theory of Inductive Inference I and II, Information and Control **7**, 1–22 (1964); 224–254 (1964).
- *Selected Works of A. N. Kolmogorov, Volume III: Information Theory and the Theory of Algorithms (Mathematics and its Applications)*, edited by A. N. Shiryayev (Kluwer, 1987).
- G.J. Chaitin, *Information, Randomness & Incompleteness*, 2nd edition (World Scientific, 1990); *Algorithmic Information Theory* (Cambridge University Press, 1987).

10. It is quite remarkable that a picture or a movie captures something about the system being photographed without having the same mechanisms as the system itself. This is the essence of description.
11. Strictly, Shannon considered the case when there is a definite way of translating between the languages.
12. The approach of considering systematically the complexity at all scales has been described in:
    - DCS, pp. 716–825.
    - Y. Bar-Yam, Complexity Rising: From Human Beings to Human Civilization, a Complexity Profile, NECSI Research Report YB-0009 (1998). http://necsi.net/projects/yaneer/Civilization.html
    - Y. Bar-Yam, Multiscale representation: Phase I, Report to the Chief of Naval Operations Strategic Studies Group (2001).
    - Y. Bar-Yam, Complexity Rising: From Human Beings to Human Civilization, a Complexity Profile, in The Implications of Complexity, edited by J. Goldstein and U. Merry, in *Encyclopedia of Life Support Systems (EOLSS)* (UNESCO EOLSS Publishers, 2002). http://www.eolss.net
    - Y. Bar-Yam, General Features of Complex Systems, in *Encyclopedia of Life Support Systems (EOLSS)* (UNESCO EOLSS Publishers, 2002). http://www.eolss.net
    - Y. Bar-Yam, Sum rule for multiscale representations of kinematic systems, Advances in Complex Systems **5**, 409–431 (2002).
    - Y. Bar-Yam, Unifying Principles in Complex Systems, in *Converging Technology (NBIC) for Improving Human Performance*, edited by M. C. Roco and W. S. Bainbridge (Kluwer, 2003), pp. 335–360.

- Y. Bar-Yam, Complexity of Military Conflict: Multiscale Complex Systems Analysis of Littoral Warfare Report to Chief of Naval Operations Strategic Studies Group (2003).
- Y. Bar-Yam, Multiscale complexity/entropy, Advances in Complex Systems **7**, 47–63 (2004).
- Y. Bar-Yam, Multiscale Variety in Complex Systems, Complexity **9:4**, 37–45 (2004).
- Y. Bar-Yam, A Mathematical Theory of Strong Emergence Using Multiscale Variety, Complexity **9:6**, 15–24 (2004).

Chapter 5 Complexity and Scale in Organizations

13. A description of the application of complexity and scale to human civilization and the inadequacy of central control for complex systems is described in the citations of Ref. 12.
14. A. Toffler, *Future Shock* (Random House, 1970).
15. The importance of being complex in a complex environment is formally stated by Ashby's "Law of Requisite Variety:"
    - W. R. Ashby, *An Introduction to Cybernetics* (Chapman and Hall, London, 1957).

    The generalization of complexity matching to multiscale analysis is described in the citations of Ref. 12.

Chapter 6 Evolution

16. Darwin's original work is still a very readable explanation of evolution and is available in various reprint editions and online:
    - C. Darwin, *The Origin of Species by Means of Natural Selection* (1859); (Reprinted by Wildside Press, 2003). http://www.literature.org/authors/darwin-charles/the-origin-of-species/
17. The Neo-Darwinian approach incorporated the role of genetics into evolution:
    - R. A. Fisher, *The Genetical Theory of Natural Selection* (Clarendon Press, 1930).
    - J. B. S. Haldane, *The Causes of Evolution* (1932) (Reprinted by Princeton University Press, 1990).
    - S. Wright, *Evolution and the Genetics of Populations. Volume 1: Genetic and Biometric Foundations* (U. Chicago Press, 1968); *Volume 2: Theory of Gene Frequencies* (U. Chicago Press, 1969); *Volume 3: Experimental Results and Evolutionary*

Deductions (U. Chicago Press, 1977); *Volume 4: Variability Within and Among Natural Populations* (U. Chicago Press, 1978).

18. The gene-centered view is described in:
    - R. Dawkins, *The Selfish Gene*, 2nd edition (Oxford University Press, 1989).
19. A formal mathematical discussion of the limitations of the gene-centered Neo-Darwinian approach is provided in:
    - DCS, pp. 604–614.
    - Y. Bar-Yam, Formalizing the gene-centered view of evolution, Advances in Complex Systems, **2**, 277–281 (2000).
    - H. Sayama, L. Kaufman and Y. Bar-Yam, Symmetry breaking and coarsening in spatially distributed evolutionary processes including sexual reproduction and disruptive selection, Physical Review **E 62**, 7065 (2000).

    Technically, the gene-centered view considers the process of evolution to be attributable to properties (i.e. fitness) that can be assigned to individual alleles. Assigning properties to an allele can be justified if we can average over the possible environments of that allele. However, if the allele happens to occur in a particular genetic environment formed by other alleles, for better or worse, this is the environment in which its success or failure will be measured by evolution. If there are several environments that the allele could appear in, we cannot average over the environments because the likelihood of a particular environment also changes from generation to generation. This change is not described by the same equation that describes the change in proportion of the allele. For example, there may be cases where the allele proportion does not change, but the proportion of each environment changes. This change in the likelihood of environments is part of evolution, not separate from it. Requiring new equations to be part of describing evolution means that we cannot assign the properties of evolutionary change to the allele.

    Conceptually speaking, it is often recognized that even in the gene-centered view, selfishness is not the whole story because the fitness of an allele has to do with how well it cooperates with other alleles that are part of the same organism. The difficulty is, however, that this property is considered a property of the allele, its "cooperativity." This makes sense up to a point. The problem is that how well one allele cooperates with other alleles also depends on the other alleles not just on itself. Only when we can average over all the possible allele combinations does the cooperativity become a property of the allele and not of the group it is part of. Thinking through an analogy to team sports, discussed in the next chapter, may be helpful to understanding this distinction. In particular, while player cooperativity might be considered as a property of a player, the degree to which a player cooperates depends on the specific teammates the player is with and therefore the average cooperativity over possible player combinations does not correspond to the actual cooperativity of the player or determine the success of a team.

20. A history of the controversy over altruism in evolution is provided in:
    - E. Sober and D. S. Wilson, *Unto Others: The Evolution and Psychology of Unselfish Behavior* (Harvard University Press, 1998).
21. A collection of articles discussing a variety of approaches to gene, organism and group selection can be found in:
    - *Genes, Organisms, Populations: Controversies Over the Units of Selection*, edited by R. N. Brandon and R. M. Burian (MIT Press, 1984).

    The scientific rejection of group selection started as a response to the pro-group selection work :
    - V. C. Wynne-Edwards, *Animal Dispersion in Relation to Social Behavior* (Hafner, 1962).

    The statement of the rejection of the idea of group selection can be found in:
    - J. Maynard-Smith, Group selection and kin selection, Nature 201, 1145–1147 (1964).
    - G. C. Williams, *Adaptation and Natural Selection: A Critique of Some Current Evolutionary Thought* (Princeton University Press, 1966).

    The biological relevance of group selection has been discussed by Sober and Wilson (Ref. 20). See also theoretical and experimental studies in:
    - M. Gilpin, *Group Selection in Predator-Prey Communities* (Princeton University Press, 1975).
    - M. Wade, Group selection, population growth rate competitive ability in the flour beetle, Tribolium spp., Ecology, **61**,

1056–1064 (1980).

22. The criticism of the gene-centered view by Sober and Lewontin can be found in:

• E. Sober and R. C. Lewontin, Artifact, cause and genic selection, Philosophy of Science **49**, 157–180 (1982).

23. Recent work has shown that models of evolution that include the separation of organisms in space (as opposed to having them all mixed together) leads quite generally to more altruistic behavior. Selfish variants create an environment in which their offspring are less successful than altruistic variants. A recent study of the origins of social behavior is:

• J. K. Werfel and Y. Bar-Yam, The evolution of reproductive restraint through social communication, Proceedings of the National Academy of Sciences (USA) **101**, 11019–11024 (2004).

Similar issues have been discussed in game theory in the context of the prisoner's dilemma:

• Robert Axelrod, *The Evolution of Cooperation* (New York: Basic Books, 1984).

A recent reference is:

• M. A. Nowak, A. Sasaki, C. Taylor and D. Fudenberg, Emergence of cooperation and evolutionary stability in finite populations, Nature **428**, 646–650 (2004).

Other complex systems issues in evolution have been discussed in:

• S. A. Kauffman, *At Home in the Universe: The Search for Laws of Self-Organization and Complexity* (Oxford University Press, 1995).

24. Recent studies of the role of space and boundaries in evolution include:

• H. Sayama, L. Kaufman and Y. Bar-Yam, Spontaneous pattern formation and genetic diversity in habitats with irregular geographical features, Conservation Biology **17**, 893–900 (2003).

• H. Sayama, L. Kaufman and Y. Bar-Yam, Symmetry breaking and coarsening in spatially distributed evolutionary processes including sexual reproduction and disruptive selection, Physical Review **E 62**, 7065 (2000).

• H. Sayama, M. A. M. de Aguiar, Y. Bar-Yam and M. Baranger,

Spontaneous pattern formation and genetic invasion in locally mating and competing populations, Phys. Rev. **E 65**, 051919 (2002).

- E. M. Rauch, H. Sayama and Y. Bar-Yam, Dynamics and genealogy of strains in spatially extended host pathogen models, Journal of Theoretical Biology **221**, 655–664 (2003).
- E. Rauch, H. Sayama and Y. Bar-Yam, Relationship between measures of fitness and time scale in evolution, Physics Review Letters **88**, 228101 (2002).
- M. A. M. de Aguiar, E. M. Rauch and Y. Bar-Yam, On the mean field approximation to a spatial host-pathogen model, Physical Review **E 67**, 047102 (2003).

Chapter 7 Competition and Cooperation

25. The idea of competition and cooperation as mutually supportive at different levels of organization can be generally understood from the ideas of group selection and more recently of multilevel selection which is discussed in the book:
    - E. Sober and D. S. Wilson, *Unto Others: The Evolution and Psychology of Unselfish Behavior* (Harvard University Press, 1998).

    A direct discussion is provided in:
    - Y. Bar-Yam, General Features of Complex Systems, in *Encyclopedia of Life Support Systems (EOLSS)* (UNESCO EOLSS Publishers, 2002). http://www.eolss.net
26. See references 18–24.
27. For example, see J. M. Smith and E. Szathmary, *The Major Transitions in Evolution* (W. H. Freeman Press, 1995).

PART II SOLVING PROBLEMS

Chapter 9 Military Warfare and Conflict

28. This chapter is based upon the article:
    - Y. Bar-Yam, Complexity of military conflict: Multiscale complex systems analysis of littoral warfare, Report to Chief of Naval Operations Strategic Studies Group (2003).

29. Discussions of the relevance of complex systems to the military can be found in the following locations:
    - Marine Corps Doctrine 6: Command & Control, U.S. Marine Corps (1996).
    - http://www.clausewitz.com/CWZHOME/Complex/CWZcomplx.htm
    - http://www.dodccrp.org/publicat.htm
    - A. Beyerchen, Clausewitz, nonlinearity and the unpredictability of war, International Security **17:3**, 59–90 (1992).
    - L. Beckerman, The Nonlinear Dynamics of War, Science Applications International Corporation (1999). http://www.belisarius.com/modern_business_strategy/beckerman/non_linear.htm
30. See, for example, A new breed of soldier, Newsweek, p. 24 (Dec. 10, 2001).
31. For discussions of the nervous system see:
    - *Neurocomputing,* edited by A. Anderson and E. Rosenfeld (MIT Press, 1988).

    For discussions of the immune system see:
    - *Design Principles of the immune system and other distributed autonomous systems*, edited by I. Cohen and L. A. Segel (Oxford University Press, 2001).
32. See:
    - S. J. A. Edwards, Swarming on the battlefield: Past, present, and future, RAND MR-1100-OSD (2000).
    - J. Arquilla and D. Ronfeldt, Swarming and the future of conflict, RAND DB-311-OSD (2000).
33. M. Van Creveld, *Command in War* (The Free Press, 1991), p. 89.
34. Ancient wars often included highly complex aspects and the equivalent of guerilla warfare is described in the bible. Even though this is not a new topic, conventional military strategy is generally designed around large scale conflict.

Chapter 10 Health Care I: The Health Care System

35. Y. Bar-Yam, Multiscale analysis of the healthcare and public health system: Organizing for achieving both effectiveness and efficiency, Report to the NECSI Health Care Inititative, NECSI Technical Report 2004-07-01 (2004).
36. World Health Report 2000 (World Health Organization, 2001).
37. See D. Altman and L. Levitt, The sad story of health care cost contain-

ment as told in one chart, Health Affairs, Web Exclusive (January 23, 2003).
38. J. Sterman, *Business Dynamics* (McGraw-Hill, 2001).
39. See Ref. 35.
40. This is the evolutionary perspective described in Chapter 15.

Chapter 11 Health Care II: Medical Errors

41. Y. Bar-Yam, System care: Multiscale analysis of medical errors—Eliminating errors and improving organizational capabilities, Report to the NECSI Health Care Inititative, NECSI Technical Report 2004-09-01 (2004).
42. Institute of Medicine, To Err is Human: Building a Safer Health System (National Academy Press, 2000).
43. A. Goldstein, Overdose kills girl at children's hospital, Washington Post (April 20, 2001).
44. M. Smith and C. Feied. http://www.necsi.org/guide/examples/er.html
45. For example, an electronic prescription system that includes automated barcode readers for checking at the bedside must allow pharmacists to change the prescription in the system.
46. The identification of physician specialty may be more or less helpful in different contexts. For example, in small hospitals physicians may be known to the pharmacist and a clear signature is all that is needed. On the other hand, for outpatient care, prescription pads could be automatically marked with physician specialty, and pharmacists are much less likely to know the physician.
A more targeted approach would focus on the specific drugs that might be confused with each other, posting a short list of the most common name-confusion culprits in pharmacies. The pharmacist would know to watch out for these drugs and when faced with a prescription for one of them, would check the clarity of the writing, or that the prescribing physician is of the right specialty. This would be an "exception handling" system: a system that deals with the cases that are likely to be problematic in a special way. However, exception handling systems place an additional large burden on the person responsible for checking for the exceptions. To be effective, every prescription must be checked against the list. If the list of exceptions includes only one or two cases, this might be reasonable. Otherwise it becomes a significantly complex task in and of itself. If the pharmacist's existing task is already highly complex, this approach would not be a good one. Moreover,

if a prescription was not written clearly, they might have to go back to the physician to ask what was desired, and such inquiries would involve additional delays and increase the burden on the physician.

47. Studies of the effect of formularies can be found in:
    - S. D. Horn, P. D. Sharkey, D. M. Tracy, C. E. Horn, B. James and F. Goodwin, Intended and unintended consequences of HMO cost-containment strategies: Results from the managed care outcomes project, American Journal of Managed Care **2**, 253–264 (1996).
    - S. D. Horn, P. D. Sharkey and C. Phillips-Harris, Formulary limitations and the elderly: Results from the managed care outcomes project, American Journal of Managed Care **4**, 1104–1113 (1998).
48. Institute of Medicine, Crossing the Quality Chasm: A New Health System for the Twenty-First Century (National Academy Press, 2001).

Chapter 12 Education I: Complexity of Learning

49. P. Senge, N. Cambron-McCabe, T. Lucas, B. Smith, J. Dutton and A. Kleiner, *Schools that Learn* (Doubleday, 2000).
50. A. Davidson, M. H. Teicher and Y. Bar-Yam, The role of environmental complexity in the well being of the elderly, Complexity and Chaos in Nursing **3**, 5 (1997).
51. R. J. Sternberg, *Thinking Styles* (Cambridge Univ. Press, 1997).
52. H. Gardner, *Frames of Mind* (Basic Books, 1983); Are there additional intelligences? in *Educational Information and Transformation*, edited by J. Kane (Prentice Hall, 1998).
53. It is interesting that some of the recommendations about interacting with children with ADD emphasize simplifying their environment as a method of amelioration. See, for example, http://add.about.com/cs/forparents/a/tipsparenting.htm

Chapter 13 Education II: The Education System

54. National Commission on Excellence in Education, U.S. Department of Education, A Nation at Risk: The imperative for educational reform (1983).
55. GDP numbers are for 2003 from:
    - World Factbook 2004 (CIA, 2004). http://www.cia.gov/cia/publications/factbook/

56. National Center for Education Statistics, Office of Educational Research and Improvement, U.S. Department Of Education, Highlights from TIMSS, The Third International Mathematics and Science Study (1999).
57. See, for example:
    - Datamonitor, Global Movies and Entertainment (2003).
    - IFPI, The Recording Industry In Numbers (2004). http://www.ifpi.org
    - Playing to win in the business of sports, The McKinsey Quarterly (2004).
58. S. Covey, *Seven Habits of Highly Effective People* (Simon & Schuster, 1990).
59. D. C. Berliner and B. J. Biddle, *The Manufactured Crisis* (Perseus Press, 1996).
60. See, for example, M. Gormley, Records show teachers cheating on tests, Associated Press (Oct. 26, 2003).
61. D. Goleman, *Emotional Intelligence* (Bantam Books, 1995).

Chapter 14 International Development

62. H. Rämi, Food aid is not development: Case studies from North and South Gondar, Report to UN Emergencies Unit for Ethiopia (2002). http://www.reliefweb.int/library/documents/2002/undpeue-eth-1jul.pdf
63. See:
    - Dams and development: A new framework for decision making, The World Commission on Dams (2000). http://www.damsreport.org/
    - Statistics on the world bank's dam portfolio, World Bank (2000). http://www.worldbank.org/html/extdr/pb/dams/factsheet.htm
64. J. D. Wolfensohn, The challenge of inclusion, World Bank (1997).
65. Many economists are familiar with the planning trap in the context of the debate between those who believe in free markets and those who believe in government planning of economic activity as was present in the Soviet Union.
66. Climate changes in Africa are described in:
    - S. E. Nicholson, Climatic and environmental change in Africa during the last two centuries. Climate Research **17**, 123–144 (2001).
    - D. Verschuren, K. R. Laird and B. F. Cumming, Rainfall and drought in equatorial east Africa during the past 1,100 years. Nature **403**, 410–414 (2000).

67. The CDF approach recognizes this, though politically based goals may compromise this recognition.

Prelude Enlightened Evolutionary Engineering

68. T. P. Hughes, *Rescuing Prometheus: Four Monumental Projects That Changed the Modern World* (Vintage Books, 2000).

Chapter 15 Enlightened Evolutionary Engineering

69. This chapter is based upon the articles:
    - Y. Bar-Yam, Enlightened evolutionary engineering—Implementation of innovation in FORCEnet, Report to Chief of Naval Operations Strategic Studies Group (2002).
    - Y. Bar-Yam, When systems engineering fails—toward complex systems engineering, International Conference on Systems, Man & Cybernetics 2003, Vol. 2, 2021–2028 (IEEE Press, 2003).

    See also:
    - Y. Bar-Yam and M. Kuras, Complex systems and evolutionary engineering, An AOC WS LSI Concept Paper, HERBB (2003).
70. There are at least three more reasons. The third reason is that unlike conventional engineering projects in which systems are built from scratch, increasingly modern engineering projects use existing "off-the-shelf" components, just as in a system in which management plays a role in determining which equipment to purchase. The fourth reason has to do with a change in the way engineers of hardware and software systems now consider these artificial entities to be interactive agents. The fifth reason is that protocols and training for human beings are also part of the system design. This further blurring of the distinction between management and engineering should not be confused with the interchangeability of human beings and computers. Human beings and computers are still good at remarkably different kinds of tasks, and the best systems recognize this and engage each in the tasks it does best.
71. Committee on Transportation and Infrastructure Computer Outages at the Federal Aviation Administration's Air Traffic Control Center in Aurora, Illinois [Field Hearing in Aurora, Illinois] hpw104-32.000 Hearing date: 09/26/1995.

72. References for the projects in the table are (J. Saltzer provided some of the references):
    - Vehicle registration, drivers license system—California Dept. of Motor Vehicles:
        - R. T. King, Jr., California DMV's computer overhaul ends up as costly ride to junk heap, Wall Street Journal, East Coast Edition, 5B (April 27, 1994).
        - J. S. Bozman, DMV disaster: California kills failed $44M project, Computerworld **28:19**, 1 (May 9, 1994).
        - C. Appleby & C. Wilder, Moving violation: state audit sheds light on California's runaway DMV network project, InformationWeek, No. 491, 17 (Sept. 5, 1994).
        - G. Webb, DMV's $44 million fiasco: how agency's massive modernization project was bungled, (California Dept of Motor Vehicles) San Jose Mercury News, 1A (July 3, 1994).
        - G. Webb, DMV-Tandem flap escalates, San Jose Mercury News, 1A (May 18, 1994).
        - M. Langberg, Obsolete computers stall DMV's future, San Jose Mercury News, 1D (May 2, 1994).
    - Automated reservations, ticketing, flight scheduling, fuel delivery, kitchens and general administration—United Air Lines:
        - A. Pantages, Snatching defeat from the jaws of victory, News Scene (monthly column), Datamation (March, 1970).
    - Statewide Automated Child Support System (SACSS)—California:
        - T. Walsh, California, Lockheed Martin part ways over disputed SACSS deal, Government Computer News State and Local (February, 1988).
        - California State Auditor/Bureau of State Audits, Health and Welfare Agency, Lockheed Martin Information Management Systems Failed To Deliver and the State Poorly Managed the Statewide Automated Child Support System, Summary of Report Number 97116 (March 1998).
    - Hotel reservations and flights—Hilton, Marriott, Budget, American Airlines:
        - E. Oz, When professional standards are lax: the CONFIRM failure and its lessons, Communications of the

ACM **37:10**, 29–36 (October, 1994).
- Advanced Logistics System—Air Force:
  - P. Ward, Congress may force end to Air force inventory project, Computerworld **IX**, 49 (December 3, 1975).
- Taurus Share trading system—British Stock Exchange:
  - H. Drummond, *Escalation in Decision-Making* (Oxford University Press, 1996).
- IRS Tax Systems Modernization projects:
  - R. Strengel, An Overtaxed IRS, Time (April 7, 1997).
- FAA Advanced Automation System:
  - U.S. House Committee on Transportation and Infrastructure, FAA Criticized for Continued Delays in Modernization of Air Traffic Control System (March 14, 2001).
- London ambulance service computer-aided dispatch system:
  - Report of the Inquiry into the London Ambulance Service, The Communications Directorate, South West Thames Regional Health Authority (February, 1993).

73. W. S. Cohen, Computer Chaos: Billions Wasted Buying Federal Computer Systems, Investigative Report, U.S. Senate, Washington, D.C. (1994).
74. Standish Group International, The CHAOS Report (1994).
75. U.S. House Committee on Transportation and Infrastructure, FAA Criticized for Continued Delays in Modernization of Air Traffic Control System (March 14, 2001).
76. The following news story reports that the FAA implemented the STARS system at Syracuse NY despite failures in meeting tests. The FAA invoked a never-before-used contract clause to force the use of this system. Due to failures of the system, flights were tracked manually.
    - J. D. Salant, Union question new traffic control, Associated Press (June, 2002).
77. Ashby's "Law of Requisite Variety":
    - W. R. Ashby, *An Introduction to Cybernetics* (Chapman and Hall, 1957).

    The generalization to multiscale analysis is described in (see also other citations in Ref. 12):
    - Y. Bar-Yam, Multiscale variety in complex systems, Complexity **9:4**, 37–45 (2004).
78. DCS, p. 756.

79. Y. Bar-Yam, Enlightened Evolutionary Engineering/Implementation of Innovation in FORCEnet, Report to Chief of Naval Operations Strategic Studies Group (2002).
80. E. Rauch, H. Sayama and Y. Bar-Yam, The role of time scale in fitness, Physical Review Letters **88**, 228101 (2002).
81. The parallel use of old systems along with new systems has appeared in a number of cases even when not planned. For example, in the implementation of the STARS system at Syracuse airport, the air traffic controllers continued to use the old system as a backup (see Ref. 76). A similar phenomenon was observed, though at least partly not intended, in the Navy's Fleet Battle Experiment Delta (FBE-D), October 1998. This experiment was conducted in conjunction with FOAL EAGLE '98 a military exercise of the Combined Forces Command Korea. In contrast to the evolutionary approach recommended in this paper, the original intention was to run a new system in parallel with the conventional one without using the new one in actual operations, but comparing their effectiveness. However, operators decided the new experimental system was better and they gravitated toward the experimental system to accomplish their tasks. As stated in a report on the experiment [Fleet Battle Experiment Quicklook Report, Maritime Battle Center, Navy Warfare Development Command, Navy War College, Newport, RI (2 November 1998) pp. 2–4]: "The original measure of effectiveness to compare current procedures with the [new system] could not be fully evaluated because operators adopted the experiment architecture in support of exercise events. This unintended use … demonstrated the [new system] value added…." Indeed, the evolutionary approach is even clearer in the following observation demonstrating the importance of real-time coexistence of innovative and conventional systems so that a new system can be tested in actual operations and the original system can be used as necessary or desirable: "As FOAL EAGLE 98 and FBE-D Delta progressed, the [new system] transitioned to full support of FOAL EAGLE. Operators used the best communication path available from the FOAL EAGLE and FBE-D capabilities."
82. D. M. Pierre, D. Goldman, Y. Bar-Yam and A. S. Perelson, Somatic evolution in the immune system: The need for germinal centers for efficient affinity maturation, Journal of Theoretical Biology **186**, 159 (1997).
83. DCS, pp. 371–419.

Prelude Global Control, Ethnic Violence and Terrorism

84. R. Frost, Mending Wall, in *Modern American Poetry*, edited by L. Untermeyer (Harcourt, Brace & Howe, 1919).
85. R. S. McNamara, *In Retrospect* (Vintage Books, 1995), p. 324.
86. S. Huntington, *The Clash of Civilizations and the Remaking of World Order* (Simon & Schuster, 1996).

Chapter 16 Global Control, Ethnic Violence and Terrorism

87. Information from interviews of Soviet émigrés and from:
    - I. Birman, *Personal Consumption in the USSR and the USA* (Palgrave Macmillan, 1989).
88. See:
    - S. Martinez, From supply push to demand pull, Amber Waves **1**, 22 (2003) http://www.ers.usda.gov/amberwaves/november03/Features/supplypushdemandpull.htm
    - J. M. Harris, Food product introductions continue to decline in 2000, FoodReview **25**, 24 (2002). http://www.ers.usda.gov/publications/FoodReview/May2002/frvol25i1e.pdf
    - Food Institute: http://www.foodinstitute.com
    - Food Marketing Institute: http://www.fmi.org
89. The World Factbook 2003 (CIA, 2003) http://www.cia.gov/cia/publications/factbook/
90. F. Zakaria, The Politics of Rage: Why Do They Hate Us? Newsweek (Oct. 15, 2001).

Chapter 17 Conclusion

91. http://www.marshallfoundation.org/about_gcm/marshall_plan.htm
92. The George G. Marshall Foundation web site states: “The Marshall Plan was a complex undertaking that is not easily described.” http://www.marshallfoundation.org/marshall_plan_examples_aid.html
93. G. C. Marshall at Harvard University on June 5, 1947; Congressional Record, 30 June 1947. http://www.marshallfoundation.org/marshall_plan_speech_harvard.html
94. http://www.nyse.com
95. D. Hock, *Birth of the Chaordic Age* (Berrett-Koehler, 1999).
96. According to widely reported statements by B. Zeitler at the Linux-

World Conference in January 2002, e.g. http://techupdate.zdnet.com/techupdate/stories/main/0,14179,2844061,00.html

97. Reports based upon estimates by Gartner, a market research firm:
    - http://news.com.com/IBM+rises,+Sun+sinks+in+server+market/2100-1010_3-5165213.html
    - http://www.midrangeserver.com/tlb/tlb060104-story01.html

INDEX

J

K

N

O

T

Knowledge Press
http://www.knowledgetoday.org

Quick Order Form

- **E-mail orders:** office@necsi.org
- **Fax orders:** 617-661-7711. Send this form.
- **Telephone orders:** Call 617-547-4100. Have your credit card ready.
- **Postal orders:** Knowledge Press, New England Complex Systems Institute, 24 Mt. Auburn St., Cambridge, MA 02138, USA. Telephone: 617-547-4100

Please send the following books. I understand that I may return any of them for a full refund—for any reason, no questions asked.

☐ Quantity: ___ *Making Things Work* (softcover) $28.95 per copy

☐ Quantity: ___ *Dynamics of Complex Systems* (hardcover) $92.00 per copy
☐ Quantity: ___ *Dynamics of Complex Systems* (softcover) $55.00 per copy

Shipping by air
U.S.: $4.00 for first book and $2.00 for each additional book.
International: $9.00 for first book and $5.00 for each additional book (estimate).

Making Things Work **organizational discount for 12 copies (10% off)**:
☐ $312 for 12 copies. Shipping: ☐ surface ($6) ☐ air(postage to be added)
☐ $624 for 24 copies Shipping: ☐ surface ($9) ☐ air(postage to be added)
☐ $936 per box (36 copies/box)
Indicate # of boxes: _____ x $936/box = Total: $_______
Shipping ☐ surface ($12/box) ☐ air (postage to be added)

☐ **Please send information on educational programs and seminars**.

Name: ______________________________
Address: ______________________________
City: ______________________ State: ______ Zip: _______
Country: _____________ E-mail address: ______________________

Payment: ☐ Check ☐ Credit card (Visa/Mastercard/AmEx only):
☐ Visa ☐ Mastercard ☐ AmEx

Credit card number: ______________________________
Name on card: ______________________ Exp. Date: ______________
Billing Address: ______________________________
City: ______________________ State: ______ Zip: _______